EARLY OHIOANS AND THEIR CRITTERS

"Painters," "Polecats,"
"Gallinippers,"
Et Cetera

Barbara Stickley Sour

HERITAGE BOOKS
2007

HERITAGE BOOKS
AN IMPRINT OF HERITAGE BOOKS, INC.

Books, CDs, and more—Worldwide

For our listing of thousands of titles see our website
at
www.HeritageBooks.com

Published 2007 by
HERITAGE BOOKS, INC.
Publishing Division
65 East Main Street
Westminster, Maryland 21157-5026

Copyright © 2004 Barbara Stickley Sour

All rights reserved. No part of this book may be reproduced or transmitted in any form or by any means, electronic or mechanical, including photocopying, recording or by any information storage and retrieval system without written permission from the author, except for the inclusion of brief quotations in a review.

International Standard Book Number: 978-0-7884-2533-1

The author thanks the following editors, publishers and/or authors who kindly gave their permission for use of their materials:

Indiana University Press for information in Douglas R. Hurt's *The Ohio Frontier*, 1990.

Larry L. Nelson for excerpts in *Jonathan Alder: His Captivity and Life With The Indians*, 1999.

Joe Besecker, Director of the Johnny Appleseed Museum and Society for material in *Johnny Appleseed: Man and Myth, 1966*.

Christopher S. Duckworth, editor of *Timeline*, an Ohio Historical Society publication, for material in the article "For Man and Beast," 1993.

Timothy J. Barbour, editor of *Echoes*, a publication of the Ohio Historical Society, for information in "Family Papers Reveal History of The Hunts," 2000.

Emily Foster, author of *The Ohio Frontier: An Anthology of Early Writings*, 1996, for material in her book.

Penquin Group, Inc., for material in Harlan Hatcher's *The Buckeye Country*, 1947.

American Heritage Pub., Inc. for the use of quotes from "Piskiou, Yaches, "Buffler, Prairie Beeves—Buffalo," by Larry Barsness, 1979.

Random House, Inc. for use of excerpts and information from the Doubleday edition of *Simon Kenton, His Life and Period 1755-1836*, Edna Kenton, 1930.

Excerpts from *The Old Wilderness Road, An American Journey*, copyright ©1968 by William O. Steele, reprinted by permission of Harcourt, Inc.

Wennawoods Publishing for material from reprint of *David Zeisberger's History of the Northern American Indians*, 1999, edited by A. Hubert & Wm. Schwarze.

Every effort was made to trace and to contact copyright holders for permission to use certain material.

Acknowledgments

There are many individuals I should like to thank for the part they played in completing this book. First and foremost is Dick Sour, my husband. He was a constant source of encouragement and support. I valued his keen observations and suggestions and marveled at his patience when I sought his advice and he had to read and reread what I'd written. With all my love, I give this book to him.

Our daughter, Dr. Leslie Sour Gertsch, drove over a thousand miles just to assist me at a critical period. A computer genius, she helped me more than she ever can imagine.

Our son, Reed, daughter-in-law, Chris, and grandson, Philip, helped me in a variety of ways from the beginning of this project to the very end.

I owe a huge debt of gratitude to the Champaign County Historical Museum, for permitting me to use their facilities. They gave me access to the books, manuscripts, newspapers, and other documents in their files. The all-volunteer staff not only was ever-willing to help but also they were a pleasant and congenial group of people with whom to work.

I'm beholden to curator, Dick Virts, whose aid and technological expertise was of tremendous assistance. He even made available some of the old and rare books in his own personal collection. I rather suspect there may have been moments when I taxed his good nature.

A very special "thank you" is due June Kiser, another Champaign County Historical Museum volunteer who came to my rescue on a sometimes daily basis. She not only was knowledgeable in her field but also was ready to offer her assistance at any time.

I find it difficult to express adequately my deep appreciation to Barbara Lehmann, a dear friend. She offered encouragement from the first moment she knew what I was doing and graciously shared her valuable and extensive library. Her depth of, and limitless, knowledge of the area in which I was researching was invaluable. She helped to correct mistakes and offered further information about things that otherwise would have gone unnoticed.

Gloria Malone at the Champaign County Library was a gem. Time after time I asked her for sources and details and she always cheerfully found the information I requested.

I also thank Barbara Culler and Emily Swickard at the Champaign County Library for their perseverance in locating pertinent information for me.

Rick Burnside was enormously helpful in permitting me to peruse a wealth of 19th century newspapers that were at his disposal. Many of the stories came from those newspapers.

I am obliged to Terry Jaworski, park manager at Cedar Bog, and to Jim Bartlett, service forester for the Ohio Division of Forestry, for their information.

There are others I wish to recognize. Kent Bauman's familiarity with growing tobacco allowed me to use the correct terminology in the chapter about snakes. W. H., (Chip) Gross, editor of Wild Ohio Magazine, and Bob Glotzhober, Curator of Natural History at the Ohio Historical Society, answered questions I posed to them about different wildlife. Tom Rieder at the Ohio Historical Society Archives Library kindly looked-up specific facts I needed. Nicole Merriman of the Government Information Services at the State Library of Ohio, Scott Caputa at the Columbus Metropolitan Library, and Phil Guss at the Ohio Agriculture Statistics Service, provided certain data I needed.

I am saddened that the late Nancy Menge will not be here to read the finished product. She spent many hours helping me and lowered my rising stress level immeasurably.

I am forever grateful to everyone I have mentioned. While working on this project I learned, once again, that the world is full of cordial and accommodating people.

                    Barbara Stickley Sour

Table of Contents

Prologue	1
Chapter 1 Snakes	3
Chapter 2 Pioneers Hated Wolves	29
Chapter 3 "Painters," and Wildcats	57
Chapter 4 Early Ohioans and "Bars"	73
Chapter 5 Deer in Ohio	99
Chapter 6 Buffalo	121
Chapter 7 Bovine	133
Chapter 8 Ohioans and Their Horses	163
Chapter 9 Pigs and More Pigs	213
Chapter 10 There Were Sheep	241
Chapter 11 Dogs, Cats, Rats, and "Polecats"	249
Chapter 12 Et Cetera Critters: Four Legged Ones	269
Chapter 13 Et Cetera Critters: Things With Wings	289
Chapter 14 An Era Ends	317
Bibliography	325

Picture Index

Page	Chapter	Topic
24	Snakes	Serpent Mound
47	Wolves	Medina Wolf Hunt
128	Buffalo	Buffalo "Green Hide"
134	Bovine	Oxen and Covered Wagon
155	Bovine	Shorthorn Bull
173	Horses	Adena
178	Horses	Emigrant's Camp
188	Horses	Stop Thief
242	Sheep	Judge Lawrence
257	Dogs, etc	Piatt Mausoleum
260	Dogs, etc	Immigrating to Ohio
324	An Era Ends	Map of Ohio

Prologue

People were settling in Ohio long before it became a state in 1803. Land was easily procured and the soil was fertile. For those interested in hunting, there seemed no end to the wildlife in the dense, verdant forests. As early as 1749, French explorer Celeron de Bienville had written about the beautiful country north of the Ohio River. Christopher Gist, a fur trader in 1751, and later a land surveyor and an agent for the Ohio Land Company, saw first-hand that the region was alive with turkeys, deer, elk, and buffalo. Trappers helped spread the word that the territory we now know as Ohio was a remarkable place. French traders and British and American soldiers were impressed with what they saw as they penetrated the untamed wilderness and realized the opportunities it offered. After fulfilling their military obligations, many of those soldiers, endowed with a spirit of adventure and a vision of the future, began the tide of western immigration. Whole families and occasionally entire communities pulled-up-stakes and traveled together. Most were poverty-stricken people who were searching for better land on which to hunt, to live, and to grow crops. The hardships they experienced along the way and after they reached their destinations might hinder but rarely stopped them. The fear of Indians was nearly constant but they were willing to take the risk. The predatory and vicious animals that sometimes plagued the newcomers did not cause them to turn back.

Much has been written about the long and difficult journeys to Ohio and the treacherous and often brutal acts on the part of both Indians and white settlers alike. Far less has been preserved about the pioneers' encounters with animals, or "critters" as they frequently were called.

The following collection chronicles not only information about the pioneers themselves but also the often dangerous and frightening predicaments with animals. Some of the tales are amusing and unusual. Not all of the situations with animals relate to wild ones; many pertain to the settlers' livestock, every now and again the very livestock they herded along with them.

Resources for these anecdotes were found in detailed journals and diaries that men, women, and older children occasionally kept. Other material was gleaned from memoirs, books, and periodicals. Abbreviated biographies in county and Ohio histories offered rewarding subject matter as well.

The narratives sometimes are gruesome and cannot qualify as pleasant reading. They are, at the same time, actual accounts of the harsh life on the Ohio frontier many generations ago. While there may be readers who disapprove of some of the things our ancestors did, we must, in all fairness, not judge them because their methods do not necessarily conform to our 21st century standards. Accept their actions for the way life was, bearing in mind on the other hand, there are episodes that do seem just a bit far-fetched. Aside from the highly unlikely possibility of confirming the true facts through consultation with the spirits of the people who experienced these events, there is no way to know what is exaggeration and what is not. The following pages are nothing more than a compilation of that which previously has been written. In many instances, direct quotations are used. The phraseology often is as interesting as the story itself.

SNAKES

Chapter 1

Prominent among critter stories are those involving venomous snakes, and most particularly, timber rattlesnakes, which were found in many parts of Ohio. Among the first to be recorded was one that took place in 1795 when a surveying crew navigating the Scioto River near the present-day site of Columbus became increasingly aware of a nauseating smell. The overpowering odor was originating from cliffs along the riverbank where an accumulation of rattlesnake skins and rattlesnake eggs were decomposing in the mid-day sun. At the same time, literally hundreds of live rattlesnakes could be seen lying on the rocks. Many more hundreds were hatching. The men found the spectacle of interest but made no attempt to land. They rapidly rowed northward.[1]

Sighting large clusters of snakes was not necessarily uncommon. During the winter months, two to three hundred and sometimes even more than one thousand, snakes habitually hibernated together in a body, only to emerge *en masse* when the weather was warm again. An abandoned well in eastern Ohio was a seasonal refuge for snakes from miles around. In the spring, harmless garter and black snakes and poisonous copperheads and rattlesnakes ranging in length from one foot to over four feet, intertwined and crawled over each other in great mounds. Settlers avoided the area whenever possible.

People also kept their distance from Rattlesnake Island, an island north of Sandusky Bay in Lake Erie. Part of a small group of islands, its sixty acres of limestone crevices and caves afforded ideal dens and were overrun with rattlesnakes. On days when they basked in the sunshine it was said not an inch of space could be seen that was not occupied by a rattlesnake. The island derived its name because of the multitude of snakes and also because the rocky ridges on the shoreline were thought to resemble the rattles of a rattlesnake.

Reverend David Zeisberger, a Moravian missionary who spent many years working among the Delaware Indians, sometimes talked about a place he said he'd never forget. He had been told that each spring when the snakes began reappearing, there were such enormous quantities of them in a certain area, one might fill several wagonloads with the loathsome creatures. Reverend Zeisberger found the claim a little difficult to believe but did agree they were present in unusually large numbers. Somewhat later in the season, however, on his second tour across the same terrain, he readily agreed with the "wagonloads" assessment. He also commented, "The air being infected with an intolerable stench . . . we held it advisable to get out of that region as soon as possible."[2]

Small, dark brown rattlesnakes, ordinarily no longer than fifteen to twenty inches, were found in swampy regions. Cedar Swamp, since renamed Cedar Bog, south of Urbana, Ohio, was one of these places. Because they resembled innocuous sticks lying on the ground, they easily could be mistaken for twigs. People residing in the vicinity of the swamps dreaded them even though their bite was not especially poisonous. Pioneers referred to them as "swamp snakes." Indians called these lilliputian rattlers "massasaugas."

During the latter part of September in the 1820s and 30s, Crawford County marshes yielded hundreds of bushels of cranberries. Some were packed in barrels and shipped to markets in the East. Many were taken to New Orleans via the Mississippi River. It usually proved a profitable cash crop for local farmers. Entire families, often coming from a great distance and camping in their wagons, assembled to pick the red berries while they were in season. The more youthful and energetic members frequently worked day and night. Always, pickers were advised to wear protection against snake bites. Although most people needed no reminding, there were those who disregarded the warning. Toward noon one day two men working near a neighbor woman heard a piercing cry and saw her throw her arms high into the air, then faint and fall to the ground. They ran to her assistance and, reviving her, learned that she had been struck by a rattlesnake. Taking a plug of tobacco from his pocket and moistening a piece, one of the men applied it to the inflamed area and repeated the procedure a number of times. The woman recovered and said "she had had enough cranberry picking that day." Other pickers found the rattlesnake and killed it, "an inevitable result of the reptile's indiscretion."[3]

Copperheads were smaller than rattlesnakes but were more feared. Rattlesnakes sometimes gave warning with their rattles before striking; not so the copperheads. The rattlesnake "was a true southerner, always living up to the laws of honor. . . . The copperhead s a wrathy little felon, whose ire is always up, and he will make at the hand or foot in the leaves or grass before he is seen."[4] It was said that the ill-tempered copperheads would not hesitate to bite a red hot coal. Children were warned that copperheads had an offensive odor much like rotting cucumbers and they must always be on the alert. The curved, hypodermic-like fangs on both rattlesnakes and copperheads made a double incision through which a poison was injected that caused an almost immediate reaction.

Reverend John Ables, of Guernsey County, related, "I was born in 1806, amongst the wolves, Indians, and snakes. . . Once I went for my uncle, Reuben Borton, through a wheat patch for a pail of water. I was terribly afraid of snakes. I stepped in my bare feet on two copperheads while going. . . . I jumped so high each time that I brought no water back. My uncle found and killed the snakes."[5]

The frightening experience of an emigrant family was narrated in an historical account of Lawrence County. Mr. and Mrs. James Kelly along with their two small children, Mr. Kelly's brother, and two hired men who were to help handle the boat, were traveling on the Ohio River. Near the Virginia (now West Virginia) shoreline they were fired upon by a party of Indians and James Kelly and his brother were killed. The two remaining men managed to steer the boat to the Ohio side of the river where one of them immediately disappeared into the forest, leaving one man, Mrs. Kelly, her five year old son, and an infant daughter to fend for themselves. Abandoning the boat, and all their provisions in their great hurry, they began walking back toward Gallipolis, the closest settlement. When they had gone but a few miles, Mrs. Kelly was bitten on the foot by a copperhead snake and could go no farther. Hiding her and her children in a clump of pawpaw trees, the man started again for Gallipolis to try to find help. Detouring far into the woods and away from the river, he hoped to mislead the Indians should they try to find the family. Four days later he staggered into the settlement where he obtained a keel boat and with a group of thirty men poled back up river to find the Kellys. They held little hope of finding them alive.

Meanwhile, Mrs. Kelly, who was in extreme agony and unable to move, sent the little boy each day to the river's edge to attempt to hail any boat that might pass. Several, indeed, did pass but fearing an Indian decoy, they ignored the child's cries for help. Daily they grew weaker from lack of food. Only a short time before the boat from Gallipolis arrived, a boat heading downriver, deciding that Indians would not station such a small child to decoy them, turned in to shore and rescued the family Later, the two boats met and a mid-river transfer was made.[6]

Simon Kenton, who guided many newcomers to Ohio, spoke of having to make camp for two weeks while one of their group recovered from the bite of a copperhead snake.

Antoine Claude Vincent lived alone in a cabin he had constructed deep in the wilderness in Scioto County. Late one evening as he stood outdoors, he felt something jab his foot. When the foot started to swell and he began to feel deathly ill, he understood what actually had happened. A snake had bitten him. He quickly left for the village where he knew there was a doctor but the foot soon was so swollen he was forced to crawl the last quarter of a mile. A written record of the episode said that Vincent lay for three weeks near the point of death and during that period experienced such violent convulsions he gnawed on and partially shredded a blanket that was used to cover him.[7]

Nearly everyone who lived where venomous snakes were prevalent made it a practice to carry a pronged stick so they could pin the snake's head to the ground while disposing of it. In gardens, snakes often lay under the sheltering lower leaves of plants where they were difficult to see. In the open prairies, the tall, coarse grass concealed them, putting both humans and livestock at risk.

A biographical sketch of an early pioneer in Greene County said that in the middle of the summer when people began harvesting their wheat, the fields would be full of rattlesnakes and copperheads. For protection against their poisonous bites men, women, and children sometimes wrapped handfuls of twisted grass around their legs, from ankles to knees.

Thurmon Necox wrote in his memoirs that one afternoon in 1801, he was hoeing corn on his farm in Meigs County. In the course of a day's time he killed five rattlesnakes that were lying under the cornstalks.[8]

Orton G. Rust, author of *History of West Central Ohio,* grew up on a farm in Clark County. He wrote that year after year he conservatively estimated an average of at least one rattlesnake per acre was killed. He vividly recalled the sensation he felt when he accidentally touched the things as they lay buried deep in the corn shocks. "Only reflex action saved him." He explained, "To find yourself soaring through the air tense with some unknown horror, your body galvanized into a lightning activity while the mind groped for the cause and only realized what was the matter when the wrist-thick rattler writhed its laggard length out from under the last wide opened corn husk, is an unforgettable experience...."[9]

William S. Madden, "went out one morning to cut a pile of hay, not far from a clump of trees opposite his cabin, in the midst of the prairie. His grass hook happened to be a sharp butcherknife, and after cutting what he supposed would make a good-sized haycock, he proceeded to gather it up in a pile, and was surprised to find thirty-seven heads of rattlesnakes, which he had cut off while cutting the grass!"[10]

Henry Howe, in his *Historical Collections of Ohio,* related the tale of two little boys in Highland County who were working in the family's tobacco patch. Six year old James and eight year old John were on their knees running their hands from the top of the pink-blossomed plants down to the lower leaves snapping off the suckers, when James was struck on the finger by a rattlesnake he had disturbed. Their parents were away from home and there was no one to tell the boys exactly what to do. The two frightened children were well aware of the possible fatal result of such a wound if left untended. James begged his brother to cut off his hand with an old hatchet he was carrying. This John refused to do. When the finger turned a blotchy-blue and swelled alarmingly, John finally consented to try to cut off the finger and begin chopping at it. The blade was dull, so very dull that it succeeded only in bruising the skin. Still, John reluctantly chopped away. Ultimately, the finger came off but in the process the whole hand permanently was damaged.[11]

A pioneer woman whose infant daughter was playing on the ground in front of the cabin "was horrified upon seeing her about to clutch a huge yellow rattlesnake. She ran and jerked away the child, and her excitement emboldened her to hunt a club with which she dispatched his snakeship."[12]

A story in 1881 told about a woman who spied a timber rattler lying in the grass and in utter panic jumped up onto a five foot high rail fence. The top rail turned, throwing her to the ground where she landed, not on the snake but, on a sharp stake that pierced "her hip, making a ghastly wound, killing her in twenty four hours. She leaves a husband and four small children, the youngest being a baby nine months old."[13]

Although the earliest settlers rarely used fences and then only to discourage livestock from trampling and eating their gardens, rail fences like the one onto which the woman leaped, were erected as time permitted and as the necessity arose. They were constructed with a series of wooden rails laid in a parallel fashion and placed one on top of the other. Owing to their zigzag appearance they occasionally were called "snake rail fences."

People were not always safe inside their own houses. Joshua Antrim's *History of Champaign and Logan Counties,* told about a family who found a rattlesnake in their children's bed. Feeling something moving at their feet one night, the terror-stricken children called to their father who discovered a rattlesnake that during the chilly night had slipped under the bedcovers where the children's body heat supplied a warm retreat. Carefully raising the covers and using a poker from the fireplace, the father gingerly lifted the serpent and tossed it into the fire.[14]

In Summit County, Mrs. John Campbell sat her baby son on the floor with a cup of milk. Later, when she returned, she was stunned to see that "close at its side was coiled a large, yellow, repulsive, rattlesnake" that had crawled through a crack in the floor. As the baby, a spoon gripped tightly in one small fist, reached for the snake, the woman shrieked in horror. As she raced into the room the rattler slid back through the crack and disappeared. Mr. Campbell came in soon afterwards and, prying up a plank in the floor, found and killed the snake.[15]

In Vinton County, a mother laid her sleeping baby in his cradle and went to the field to help her husband. An Irishman who was employed by the family went inside to replenish his supply of chewing tobacco and saw a rattlesnake lying by the cradle. Rushing to the field, he shouted, "Oh, mon! come quick; the devil he is in the house!" At their entrance, the snake crawled under the cradle where the baby's father yanked it out and killed it with a club.[16]

A family near Warren, Ohio, was eager to occupy their partially constructed log cabin and decided, completed or not, they would make their beds and sleep inside the structure. "All went charmingly, until one morning the mother, in making up the bed in which they had slept, in drawing off the feather tick in order to shake up the straw tick, discovered to her consternation and terror a large rattlesnake gliding away between the logs, which was supposed to have ensconced itself between the two ticks the day before; and during the night had remained so quietly still as not to have disturbed its bed fellows."[17]

The fear of snakes had an impact on the site of a popular camp meeting in Champaign County. Camp meetings were an important part of Ohioans' lives from the early-to-late 19th century. People looked forward to these outdoor religious gatherings with much anticipation. Families would stay for days at a time making new acquaintances, racing and trading horses, and listening to the messages of the evangelistic-style circuit riders.

For more than five years an annual camp meeting was held in a pleasant grove of trees near Mad River, northeast of Westville, Ohio. People had grown accustomed to camping in their wagons and in the primitive log "tents" some erected under the shade of the trees or along the bank of the river. One summer, however, rattlesnakes were sighted and a small boy was reported to have been bitten. Dissension had been growing for some time among a few of the camp-goers and it was this faction who was advocating moving to a different vicinity. They began calling the Mad River location "rattlesnake camp meeting." Almost immediately, attendance fell to an unprecedented low and not long thereafter camp meetings at Mad River were discontinued. Land southeast of Urbana was offered—land without rattlesnakes people were assured. A quarter of a century later the new location became one of the most highly recognized and well-attended camp meetings in the nation.[18]

There was a region in Summit County where "rattlesnakes were very numerous and a great pest to the first settlers of Stow township." On sunny days in the spring when the lethargic creatures began emerging from their winter dens deep in the rocky crevices, men in the neighborhood took turns watching for them and killing them. A Mr. Baker was taking his turn one warm Sunday morning while church services were being conducted in a nearby house. "With a long pole prepared for the purpose . . . , " he "killed sixty-five of the venomous reptiles." His young son, in great excitement, ran to the meetinghouse, shouting at the top of his voice, "Oh, Dad's killed a pile of snakes!" Worship abruptly ceased and the whole congregation rushed to see and to "give their thanks for the victory over the ancient enemy of mankind."[19]

Few pioneers were tolerant of poisonous snakes. They would rid themselves of the menace at every chance. In a rocky area near Granville, Ohio, inhabitants of the tiny settlement discovered a great mass of rattlesnakes, black snakes, and copperheads coiled together in an immense heap. More snakes were observed crawling under the rocks. Dividing into two teams and arming themselves with heavy clubs, the men organized a gigantic snake hunt. It was agreed that the losing side would provide whiskey for all the hunters. At day's end, the winning side was said to have disposed of more than three hundred rattlesnakes and copperheads. The losing team good-naturedly supplied several gallons of whiskey. Both groups participated in consuming it.[20]

Close to five hundred rattlesnakes were killed in 1798 in southern Ohio by a determined party of men who descended onto a ledge of rocks that was overrun by snakes. According to the men, a majority of the reptiles were larger than a man's leg below the calf and longer than five feet in length. One of the men who joined in the hunt returned to the rock ledge by himself the following day armed with a long-handled, knife-like weapon. He positioned himself in a butternut tree that extended out over a chasm in the rock. Eight feet below swarmed the writhing rattlesnakes. Before he could use his improvised weapon, he lost his balance and slipped. Terrified, he reached out blindly and grasped a small branch in the tree to prevent himself from plunging headlong into the snakes' den. "I could not have gotten out had there been no snakes, the rocks on all sides being nearly perpendicular. It was a merciful and providential escape," he wrote.[21]

During the spring and summer months of 1801, the founder of Springfield, Ohio, James Demint and his wife, often could be seen roaming among the rocks and cliffs along Buck Creek smoking timber rattlers from their lairs and clubbing them to death. An early history of the county estimated that the couple "killed ninety snakes one spring, an unbroken local record."[22]

James Swisher, whose home was near North Lewisburg, told about his experiences following the Civil War when he traveled and worked in the West. For a short time he stayed in Rattlesnake Gulch, California, which he acknowledged was appropriately named and where he saw literally thousands of rattlesnakes; however, he continued, a two hundred foot high slope near the town of Green River, Utah, had many, many more. He never ventured far from camp before cutting a heavy club. One morning he counted twenty on the path where he was walking. Quickly changing his mind, he turned and walked in a different direction. Some of the rattlesnakes he saw on the slope that day were seven feet long.[23]

There were people who lost their fear of rattlesnakes and took dangerous risks. It was thought to be fine sport by some men to pin a snake to the ground and then seize it by its tail. Removing the stick and giving a sudden backward jerk with its tail, the snake's spine was broken. A young man who was engaged and soon-to-be-married tried the daring feat with an extra large rattler only to have it slip away from his grasp and crawl into a tree stump. It swiftly coiled and struck him on the wrist. He died a short time later.[24]

Other impetuous youths made a contest of flinging poisonous snakes against trees before they had time to strike or curl around their arms.

Snakes could coil very quickly. A respected minister once told of watching a hawk swoop down and clutch a snake. Before the bird could eat the snake, the snake wrapped itself around the bird and killed it.

An unusual method of disposing of a rattlesnake was reported in the August 9, 1877, *The New Era*, a St. Paris, Ohio, newspaper. While preparing dinner, a woman, to her horror, discovered a rattlesnake lying in the oven of her stove.

Fearing it might escape, she quickly slammed the oven door shut, kindled a fire, and baked it.

Daniel Constable, one of two brothers who came from England to travel and to see the United States, spent much of his time in Ohio. Usually, he traveled on-foot. Not only did he keep a daily journal but he also managed to keep up a lively correspondence with his parents and siblings. In an 1822 letter to his parents, he wrote: "I forgot to tell you I had a very perfect fine R Snake skin, bread and brought up on the Volney's farm 5 ft 9 In now dry, have his jaws & teeth, it was the smallest of three that were killed soon after I got there, it is extended upon my wall."[25]

In 1755, Delaware Indians captured eighteen year old James Smith and adopted him into their tribe. He was treated with much kindness and affection and during the five years he spent with them became familiar with some of their myths and folklore. He learned that rattlesnakes played an important role. Many of the Indians would not kill rattlers, believing the creatures possessed supernatural powers. Instead, they chose an almost-protective relationship with them. In the winter months, James' "family" often camped in what is now Mahoning County where they set traps for raccoons and hunted for deer. They theorized that rattlesnakes turned into raccoons every fall and, in the spring, became rattlesnakes again. They based this premise on their own observation for in cold weather the traps they set normally were full of raccoons; however, with the advent of spring, the traps no longer held raccoons but rattlesnakes. This led them to the conclusion rattlesnakes were transmigratory.[26] Although the Indians rarely killed any of the rattlesnakes, they were not adverse to dipping the tips of their arrows in the venom.

A few Indian tribes thought the number of rattles on the end of a rattlesnake's tail indicated how many individuals the snake had killed.

Ohio Indians generally are given credit for teaching the pioneers in this part of the country about common plants growing in the surrounding forest that were effective antidotes for snake bites. Interestingly enough, these plants—for example, Virginia snakeroot, Seneca snakeroot, and white snakeroot—were said to have derived the name "snake" primarily because their roots bore a resemblance to snakes and not necessarily because of their medicinal value. The stalks of the snakeroot plants, which grew approximately nine inches tall, had spiraling yellow roots about the diameter of small knitting needles.

Some tribes contended that an orchid they referred to as "rattlesnake plantain" possessed healing qualities for a snake bite because its leaves were the same coloration as a serpent's skin.

Indians recommended that a person who had been bitten by a rattlesnake chew the stalks of one of these plants, swallow the juice, and lay the remaining mouthful of now-moist, masticated stems on the inflamed surface. This induced vomiting which helped to rid the patient's system of the poison. If done quickly enough, any serious consequences could be averted.[27]

Native Americans were known to use tobacco (Kin-nec-ka-neek) as a rattlesnake repellent and frequently spread it liberally on the ground when they made camp for the night. They smoked it in their pipes before going to sleep and sometimes tied bits of it to their ankles in the belief it would chase away any snakes.[28]

Dr. Conant, a physician near Chillicothe once told about an adolescent boy who had been bitten by a timber rattler. When he arrived, the boy was lying in the yard "horribly swollen, as spotted as a leopard and in great agony" and he concluded it was too late to save him. An Indian who was observing the scene with deep interest informed the doctor, with considerable contempt, that one seldom heard of the death of an Indian from venomous snakebites. He told the boy's parents that for a gallon of whiskey, he could heal their son. The parents, of course, agreed. Carefully concealing from the doctor which plants he was using, the Indian treated the wound and the boy survived. The grateful parents gave the Indian his gallon of whiskey. Dr. Conant gave the Indian three gallons of whiskey to tell him how he'd cured the boy.[29]

Another 19th century doctor was called upon to administer to a small boy who had been bitten by a rattlesnake. He had nothing with him that would help the child so he seized the snake, which had been killed and tossed aside, and cut a two inch piece from the middle of its body. Splitting it, he bound the flesh side to the wound. Several hours later, the poultice was removed and the boy appeared as healthy as ever.[30]

A man who frequently hunted in the woods had on countless occasions cause to treat venomous snakebites. His cure was very much like the old doctor's. He bound "the liver and guts of the snake to the bite."[31]

Favorite remedies to counteract poisonous snake bites often were passed from person to person. One directed the user to gather "cucold bur leaves and bruise them, put them in sweet milk, strain and drink the same." Of a more questionable nature was the advice to bind garlic to the bottom of the afflicted person's feet.[32]

Possibly the most bizarre treatment on record originated with Jesuit priests who were among the first white people in the largely unsettled western lands. "The most common and efficacious remedy consists of securing the head of the snake between two sticks, keeping the head in such a position that the snake cannot bite. Then the tail is held firmly and stretched out so that the snake cannot coil. The victim of the snakebite then bites the snake. At this point, something truly remarkable happens. The patient does not swell, but the snake does, monstrously so, until it bursts."[33]

Reverend David Zeisberger said it had been reported to him by both Native Americans and whites that a rattlesnake, if irritated in some way but unable to locate the source, would sink its fangs into its own body. As a result, it would swell and die in a few hours.[34]

Martin Root and two companions near Worthington, Ohio, demonstrated that a human's bite was as venomous as a snake's. They spied a rattlesnake and while one of the young men anchored it to the ground, the others forced a well-chewed quid of tobacco into its mouth. They then turned the snake loose. It "did not crawl more than his length before it convulsed, swelled up and died, poisoned to death by the virus from the mouth of one of the lords of creation."[35]

Horses and cattle were not immune to poisonous snakebites and if not given prompt attention could die in a short time. Usually they were struck on their feet or, if lying down, on their nose. With proper care, they might recover in twenty-four hours, far more quickly than most human beings.

One man's formula for healing snakebites on livestock was to place a poultice of snakeroot on the wound—when he could find it—otherwise, he made a bandage of bear's oil. If bear's oil was not available, he used the fat from a piece of meat in order to draw out the poison and relieve the swelling. The greatest challenge often was keeping the animal quiet.

A. W. Chase, M.D., offered a treatment for horses and cattle in a book he published. "All that is necessary to do is to drive them into a mud-hole and keep them there for a few hours; if upon the nose, bind the mud upon the place in such a manner as not to interfere with their breathing. And I am perfectly satisfied that soft clay mud would be an excellent application to snake bites on persons, for I know it to draw out the poisoning from ivy, and have been assured that it has done the same for snake bites, of persons as well as for cattle."[36]

Ordinarily, rattlesnakes were regarded with abhorrence but there were occasional instances when they were perceived as a blessing. An 18th century team of surveyors marking land claims recorded their experience when their food supply began to run perilously low. They finally were reduced to consuming wild berries and broiled rattlesnakes, all washed down with rum. The berries, they contended, gave them indigestion, the water in the streams gave them dysentery, and the clouds of mosquitoes gave them malaria. Only the rum and the rattlesnakes gave them no discomfort.[37]

In 1796, Joshua Stow was a flagman with the first party of surveyors in the Western Reserve and, as flagman, preceded the others. He frequently encountered rattlesnakes which he would kill and drape over his shoulders. There were few nights when he did not have from two to eight of the sinister things and he presented a grotesque appearance as he strode into camp with them swinging rhythmically at each step he took. The snakes were dressed, cooked and eaten, offering an almost-welcome respite from the usual fare of salted meat.[38]

A team of four surveyors in Washington County employed hunters to keep them supplied with meat. One day the hunters were unable to find any wild game but killed a large rattlesnake which they cut into slices and broiled over the coals in the campfire. Although one of the surveyors refused even to try it, the remainder of the group found "the flesh was sweet and good."[39]

Although he hated snakes, Elisha Fees, a soldier during the French and Indian War, admitted to being everlastingly grateful for a rattlesnake he credited with helping to save his life. While on maneuvers, his detachment was ambushed by Indians and he received a musket ball in his thigh. In a desperate effort to hide from the Indians, he managed in the confusion to drag himself to a log lying nearby. Sliding under it as far as he could, he succeeded in escaping their detection. Barely able to move and dizzy from pain and the loss of blood, he lay for six long days. Hunger and thirst became a serious concern and when, on the third day, he killed a rattlesnake that ventured too close, he skinned it and devoured it raw. On the sixth day, he crawled from under the log. By a stroke of good fortune, he patiently coaxed a horse the Indians had not captured and was running loose, to stand still long enough for him to reach its reins and pull himself to a stump. Wincing and with much difficulty, he climbed into the saddle and rode to the fort "where he was joyfully received and his wounds kindly cared for."[40]

Most people had an innate fear of snakes, even when they expressed it in a jovial manner. Philip Kenton, a nephew of the famous frontiersman, Simon Kenton, was asked to give an impromptu speech one evening in Cincinnati. He quipped, "I cannot make a speech, but can say I always hated snakes and loved the women."[41]

At the opposite end of the scale was the legendary John Chapman, better known as Johnny Appleseed. He was renowned for his unswerving kindness to people and animals alike and was a familiar figure in the Ohio landscape. He was considered somewhat eccentric but genuinely was loved by all who knew him. He fervently believed that no person had the right to cause the death of another living creature and when, in a moment of anger, he once killed a rattlesnake after it bit him, he felt remorse.[42]

An acquaintance laughingly said the skin on the bottom of Johnny's feet was so thick, "any attempt to bite through such hide as his surely would kill a rattlesnake."[43]

Occasionally, when Johnny planned to remain in a community long enough to plant a large nursery of apple seedlings, he would ask boys in the neighborhood to help him construct a small cabin. It ordinarily consisted of little more than a framework of poles covered with a bark roof.

In Richland County, two boys arrived at the campsite to assist him in erecting such a cabin. Before retiring for the night, one of them pulled a log closer to the fire. The movement roused a rattlesnake and when it coiled to strike, the boy leaped up to kill it. Johnny, however, sternly rebuked him, reminding him that it was he who had bothered the snake.[44]

Episodes with snakes frequently were printed in local newspapers. An 1879 St. Paris, Ohio, newspaper said, "One day last week Laura Bell, daughter of John W. Kizer, of Harrison township, was bit by a rattlesnake. The reptile struck the girl in the heel of her foot, and in a short time the foot and ankle became swollen to an alarming extent. Mr. Kizer immediately applied to the wound an unfailing remedy, the snakeroot, and in twenty-four hours all danger was over."[45] An 1880, *St. Paris Examiner* read, "A tramp reported having killed a snake on the railroad track between Springfield and Urbana the day before yesterday and his snakeship was as lively as is ordinary in July."[46] A Mechanicsburg news item listed several incidents with rattlesnakes. Joseph Metzger killed a rattlesnake in his garden. Nate Byers killed one in his "celery field" and Walter Kennedy, an operator at Catawba Station "was entering his stable the other day when a rattler came tumbling down past his head from someplace above. The venomous reptiles are evidently numerous this year."[47] Another newspaper reported that a rattlesnake with seven rattles and a button had been killed.[48] According to the April 13, 1888, issue of the same paper, a gentleman by the name of George Hatton killed seventeen rattlesnakes and one of them was a rattlesnake with seventeen rattles.

An 1878 paper carried an article about a freak accident. A doctor who acquired two poisonous snakes intended giving them to a zoological garden. He secured them in a glass jar covered with wire mesh and carefully placed them on a table in a remote corner of his office. That same evening, accompanied by his daughter, he returned to the office to fill a prescription. As he lit the lamp, an owl flew in through the door and frantically began flying around the room. It knocked over the lamp, plunging the room into darkness. The owl careened into a row of bottles sitting on a shelf. Glass could be heard crashing to the floor. Suddenly, the daughter began screaming. The snakes had escaped from the jar and one had struck her on the leg. Every antidote the father had available was administered but nothing was successful. She died a few hours later. "The corpse of the young lady had swollen to enormous dimensions, while her complexion was in harmony with the spots and general coloring of the snake." Neighbors who were alerted by the turmoil arrived and helped to recapture the two snakes. "The largest of the snakes measured four and one half feet and the smaller four feet." An examination of the doctor's boots revealed that he, too, had been struck but the thick leather protected him.[49]

Zoological gardens, museums of natural wonders, circuses, and traveling medicine shows, aspiring to attract crowds of people, sought the largest, the smallest, the ugliest, and the most unusual creatures they could find. In 1813, a visitor to one of the museums of natural wonders saw wonderful exhibits and was entranced with all that he saw. He acknowledged, however, that a certain snake commanded his greatest interest.

In metropolitan cities, zoological gardens were fashionable and drew unprecedented crowds. A newspaper carried the following remarkable story about one of the prestigious zoological gardens.

One of their prize reptiles became ill and nothing they did for it or gave it to eat seemed to be of any value. It steadily grew worse. In an effort to make the snake more comfortable, a blanket was laid in the cage but sometime during the night the snake swallowed it. His body distended accordingly and he became even more ill. For over a month the snake's health deteriorated, when one morning the blanket was discovered on the floor of the cage. Almost immediately after disgorging it, the snake began to improve. Officials determined that the blanket had served as a cathartic. The snake began growing so rapidly it was necessary to move it to a larger cage. The blanket, except for a slight discoloration, showed no visible damage and was "packed away in the superintendent's private office."[50]

There seldom has been a period in Ohio history when a "big snake" story has not been in circulation somewhere. Every now and then one surfaced in Johnson township, Champaign County. Johnson township at one time primarily was a thickly timbered, hilly region with numerous creeks and marshy areas. Reports of a black snake fifty feet long, some insisted it was closer to sixty feet long, and "as round as a dinner plate," persisted for decades. Yet today, the story occasionally can be heard.

The fact that there were scores of snakes in Ohio was accepted as a part of everyday life and rarely did it dishearten or intimidate those who came to settle here. Poisonous snakes merely were one of the many obstacles to overcome. A woman in Vinton County who wrote about the difficult life the newcomers experienced, remarked that "A wilder country than this in the early days it would be hard to imagine, with its great systems of rocks and intermingled forests. Indians, wolves, wild game and snakes were more numerous than interesting."[51]

Lovell Calkins, who came to Ohio in the early 1800s, was so optimistic about the prospects he saw that he described the land as a "second Eden (not even lacking the serpents), and induced others to emigrate to its delectable fields."[52]

As the earliest pioneers and succeeding generations steadily cleared the wilderness, however, and civilization intruded, the snake population dwindled. Today, more than a century later, copperheads and timber rattlers exist only in a few rocky, remote areas in Ohio.

Massasauga snakes still can be found in their swampy world but decades of systematic draining have resulted in their being placed on the endangered list.

Garter snakes and black snakes also are seen less and less frequently. There are, however, youngsters today who, much like Jefferson County's thirteen-year-old John McCracken and his friends circa 1830, keep garter and black snakes for pets and periodically "bring them out, sit on their doorstep and let them crawl over them."[53]

Probably the most outstanding, certainly the most publicized reminder of snakes in the State of Ohio, is Serpent Mound, a Native American earthworks in Adams County. Shaped like a 1,348 foot long snake in the act of swallowing an egg, it was constructed long before the white man arrived in this part of the country. Thought at one time to have been an object of worship, its purpose still remains unclear although it has been excavated and proven not to be a burial mound. Now owned by the State, this unique representation of a snake draws visitors from all over the world.

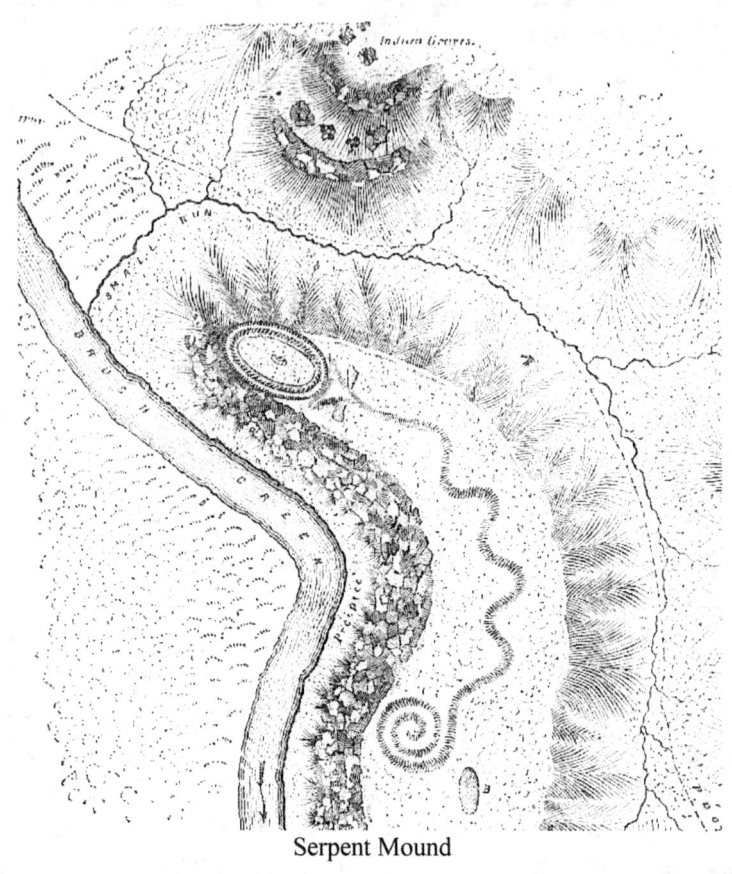
Serpent Mound

Reprinted from Howe, *Historical Collections of Ohio*, Vol. 1, 1896, p 232.

Chapter 1: SNAKES

1. Betty Garrett, *Columbus, America's Crossroads* (Tulsa, OK: Continental Heritage Press, 1980), p 22.
2. Rev. David Zeisberger, *David Zeisberger's History of the Northern American Indians,* ed. Archer Hulbert & William Schwarze (1910; reprint, Lewisburg, PA: Wennawoods Publishing, 1999), p 71.
3. Henry Howe, *Historical Collections of Ohio,* vol. 1 (Norwalk, Ohio: Laning Printing Co.,1896), p 488.
4. ——, vol. 2 (Cincinnati: Krehbiel & Co., 1904), p 79.
5. ——, vol. 1 (Norwalk, Ohio: Laning Printing Co., 1896), p 733.
6. ——, vol. 2 (Cincinnati: Krehbiel & Co., 1904), pp 56- 57.
7. ——, p 566.
8. Stillman C. Larkin, *Pioneer History of Meigs County* (Columbus: Berlin Printing Co., 1908), p 64.
9. Orton Rust, *History of West Central Ohio,* vol. 1 (Indianapolis: Historical Pub. Co., 1934), pp 356- 357.
10. W. H. Beers, *History of Champaign County* (Chicago: W.H. Beers & Co., 1881),p 289-290.
11. Henry Howe, *Historical Collections of Ohio,* vol. 1 (Norwalk, Ohio: Laning Printing Co., 1896), p 923.
12. Joshua Antrim, *History of Champaign and Logan Counties, From Their First Settlement* (Bellefontaine, Ohio: Press Printing Co., 1872), p 27.
13. *St. Paris (Ohio) New Era,* July 16, 1881.
14. Joshua Antrim, *History of Champaign and Logan Counties, From Their First Settlement* (Bellefontaine, Ohio: Press Printing Co., 1872), p 241.
15. Henry Howe, *Historical Collections of Ohio,* vol. 2 (Cincinnati: Krehbiel & Co., 1904), p 656.
16. ——, p 736.
17. Joshua Antrim, *History of Champaign and Logan Counties, From Their First Settlement* (Bellefontaine, Ohio: Press Printing Co., 1872), p 26.
18. William J. Knight, "Some of the Early Campmeetings in Champaign County," unpublished manuscript (circa 1906).
19. Henry Howe, *Historical Collections of Ohio,* vol.2 (Cincinnati: Krehbiel & Co., 1904), p 656.
20. ——, p 77, 79.
21. ——, (Cincinnati: Derby, Bradley & Co., 1848), p 480.

22. Orton G. Rust, *History of West Central Ohio* (Indianapolis: Historical Pub. Co., 1934), p 356.
23. James Swisher, *How I Know* (Cincinnati: Press of Jones Bros., 1880), p 229.
24. Joshua Antrim, *History of Champaign and Logan Counties, From Their First Settlement* (Bellefontaine, Ohio: Press Printing Co., 1872), p 27.
25. J. Brian Jenkins, *Citizen Daniel (1775-1835) and The Call of America* (Hartford, CT: Aardvark Editorial Services, 2000), p 270.
26. Henry Howe, *Historical Collections of Ohio*, vol. 2 (Cincinnati: Krehbiel & Co., 1904), p 588.
27. ——, p 494.
28. Manny Rubio, *Rattlesnakes, Portrait of a Predator* (Washington, D.C.: Smithsonian Institution, 1998), p 172.
29. Henry Howe, *Historical Collections of Ohio* (Cincinnati: Derby, Bradley & Co., 1848), p 207.
30. A.W. Chase, *Dr. Chase's Recipes; or, Information for Everybody: An Invaluable Collection of About Eight Hundred Practical Recipes, For Merchants, Grocers, Saloon-Keepers, Physicians, Druggists, Tanners, Shoe Makers, Harness Makers, Painters, Jewelers, Blacksmiths, Tinners, Gunsmiths, Farriers, Barbers, Have Been Added A Rational Treatment of Pleurisy, Inflammation of the Lungs, and Other Inflammatory Diseases, and also for General Female Debility and Irregularities: All Arranged in Their Appropriate Departments* (Detroit: F. B. Dickerson Co., 1867), p 154.
31. R.S. Banta, *The Ohio* (New York: Rinehart & Co., 1949), p 399.
32. ——.
33. Robert Peck, *Land of The Eagle* (New York: Summit Books, 1990), p 195.
34. Rev. David Zeisberger, *David Zeisberger's History of the Northern American Indians*, ed. Archer Hulbert & William Schwarze (1910; reprint, Lewisburg, PA: Wennawoods Publishing, 1999), p 72.
35. Henry Howe, *Historical Collections of Ohio* (Cincinnati: Derby, Bradley & Co., 1848), p 297.
36. A. W. Chase, *Dr. Chase's Recipes . . .* (Detroit: F. B. Dickerson Co., 1867), p 154.
37. R. Douglas Hurt, *The Ohio Frontier* (Bloomington & Indianapolis: Indiana University Press, 1996), p 198.
38. Henry Howe, *Historical Collections of Ohio*, vol. 2 (Cincinnati: Krehbiel & Co., 1904), p 655.
39. ——vol. 1 (Cincinnati: Derby, Bradley & Co., 1896), p 820.

40. W. H. Beers, *History of Brown County* (Chicago: W. H. Beers & Co., 1883), p 15.
41. ——, *History of Champaign County* (Chicago: W. H. Beers & Co., 1881), p 629.
42. A. Banning Norton, *History of Knox County* (Columbus, Ohio: R. Nevins Printer, 1886), p 130.
43. Robert Price, *Johnny Appleseed: Man and Myth* (Urbana, Ohio: Urbana University, 2001), p 169.
44. Harlan Hatcher, Robert Price, Florence Murdoch, John W. Stockwell, Ophia D. Smith, and Leslie Marshall, *Johnny Appleseed: A Voice in The Wilderness*, 6th ed. (Cincinnati: Johnny Appleseed Memorial Library, 1966), p 54.
45. *St. Paris (Ohio) New Era*, August 7, 1879.
46. *St. Paris (Ohio) Examiner*, January 22, 1880.
47. *Urbana (Ohio) Citizen and Gazette*, October 4, 1888.
48. ——, November 30, 1882.
49. ——, October 17, 1878.
50. *St. Paris (Ohio) New Era*, Aug. 22, 1880.
51. Henry Howe, *Historical Collections of Ohio*, vol. 2 (Cincinnati: Krehbiel & Co., 1904), p 736.
52. *Ohio Historical Review* (Columbus, Ohio: pub./ed., Ralph S. Estes, n. d.).
53. Henry Howe, *Historical Collections of Ohio,* vol. 1 (Norwalk, Ohio: Laning Printing Co.,1896), p 976.

PIONEERS HATED WOLVES

Chapter 2

Critters other than poisonous snakes proved to be a menace to early pioneers. They feared and hated wolves. Although there are many accounts of unprovoked attacks on human beings by wolves, it has been said no death in Ohio definitely could be attributed to them. People—and usually it was a lone person—who were pursued by wolves managed to escape or had the capability to thwart the attack.

An 1881 history included a brief item regarding the death of Edward W. Pierce, an attorney. In 1816, his body was found in the forest and although it had been badly torn by wolves, it never was determined if the wolves had killed and desecrated the remains or if the man perished and later was found by the wolves.[1]

William Dean Howells, in his book, *Stories of Ohio,* repeated an episode about Daniel Boone who, with a friend in 1769, was taken prisoner by Indians. Both men escaped and a short time later were joined by Daniel's brother and his brother's neighbor. Boone's friend, however, was tracked, killed and scalped by the Indians. His brother's neighbor became hopelessly lost in the wilderness and was said to have been devoured by wolves.[2]

Colonel Robert Patterson, who later settled not far from Dayton, in Montgomery County, Ohio, told of an unforgettable experience when Indians attacked his party and the bodies of two men who died as a result of that attack were consumed by wolves. Patterson and six companions were traveling up the Ohio River toward Pittsburgh in the fall of 1776 when they were surprised at their campsite by Indians near present-day Athens County. Those not killed were badly wounded. One man, less seriously injured than the rest, went for help to a settlement approximately one hundred miles upriver. The remaining three struggled, little by little, to an outcrop of rock that offered some slight shelter from the weather. Barely alive, they subsisted for more than a week on a diet of water and pawpaws. During the long nights, the men could hear the wolves howling at the location where their friends lay exposed on the ground and surmised what was occurring but were powerless to prevent it. Although the men finally gave up hope of being rescued, help did arrive. After cleaning and dressing the men's wounds, the group searched for the bodies of the two dead men. All that remained were the bones. They were carefully assembled, placed together in a grave, and given a proper burial.[3]

Nighttime travelers usually made it a habit to carry a lantern or a torch of some type for predators ordinarily were terrified of fire. If hunters felt confident there were no Indians in the vicinity, they occasionally kept a string of campfires burning in a ring around them when they stopped for the night.

Joseph C. Vance once wrote that as a boy, he made an overnight trip to the salt licks. He had to keep a fire burning all through the night and carefully watch over his team tethered nearby because the camp was "so beset by wolves."[4] Horses and oxen could suffer serious bites from wolves.

Jonathan Alder, a white man who had been captured when he was a young boy and reared as a member of a Mingo Indian tribe in Ohio, could attest to the fact wolves were dangerous. Many years after he had located his biological family, married a white woman, and reared children, he dictated to his eldest son, Henry, his memories of life among the Indians. Relating one of his own experiences with wolves, he prefaced the story by saying "The wolf was the most likely of any to attack a person."[5]

Not quite twelve years old, Jonathan had accompanied one of the Indians on a hunting expedition. The boy was to tend camp while the older man hunted. When, a little after dark, the Indian had not yet returned, wolves began to howl. They boldly came close to the camp. Jonathan could hear their teeth snapping. "Occasionally one would come so near that I could see the shine of his eyes." He was certain he was about to be devoured. He shot all of his arrows and didn't know whether he hit one or not. Instead of adding fuel to the fire to make it burn more brightly and frighten the animals away, he tried to extinguish it. The wolves came closer. Dismantling the tent and laying it on the ground, the frightened youngster rolled himself in it, hoping it would protect him from the wolves and possibly, just possibly, his companion would return in time to save him. Not long thereafter, Jonathan was relieved to hear rapid footsteps on the hard crust of the snow and reasoned that aid was on the way. With his gun, the advancing Indian shot into the pack of wolves and from the quickly rekindled fire, "threw some firebrands at them and they soon left."[6]

Unlike deer, bears, and turkeys, wolves were not a desirable food for human beings. Only people near starvation ever considered eating their flesh. Even dogs ordinarily would not eat the meat.

Jonathan Alder remembered a period when, as a young man, he had spent so long setting-up a winter camp he neglected to replenish his food supply. Because in his haste he foolishly ran outdoors in his bare feet one day to chase an otter, he froze his feet. A few days later, he stepped on a honey-locust thorn that pierced his foot. As a result, he was housebound for several weeks. Although he ate sparingly, the food disappeared surprisingly rapidly. Two fifteen year old Indian boys chanced to stop at the camp and did their best to help. Both were fairly competent hunters and daily looked for game. They were, however, unable to kill a single animal for with every step they took, the crust on top of the snow broke through and the crunching sound alerted any game. The boys could not get close enough to fire a shot.

On the third day with nothing to eat, Jonathan sat in the doorway of the cabin staring despondently toward the creek where he had seen and chased the otter when he noticed a wolf. Grabbing his gun, he shot it. When the boys returned, dejected, hungry, and empty-handed, he told them about the wolf. Although revolted at the thought of eating it, they retrieved the creature, skinned it, and roasted the hind quarters. The flavor was so repugnant they chewed but could not force themselves to swallow it.

It was through the efforts of another wolf, nevertheless, that the three did obtain a deer. Later the same day, Jonathan heard a deer bleating in the not-too-far distance and comprehended what was occurring; a wolf was running-down a deer. The boys reached for their rifles but Jonathan restrained them, explaining that they should wait until the wolf had killed the deer. Otherwise, he said, the wolf would become aware of their presence and they would succeed only in frightening both it and the deer. The boys tarried as Jonathan suggested and a bit later, came back triumphantly carrying a fine buck deer.[7]

Winter was a difficult time for pioneers and wolves alike. During a particularly severe storm in northeastern Ohio, in 1801, wolves attacked a herd of cattle. According to some of the inhabitants who witnessed the onslaught but were unable to reach the herd, the cattle huddled together while some of the oxen and stronger animals tried to defend them, running back and forth, bellowing and, with horns lowered, charging the wolves. Despite their valiant effort and protective measures, a number of the weakest cattle were horribly wounded.

In 1830, there was a January flood in Putnam County when wolves, working in a pack, herded deer from the dry ridges where they had taken refuge, onto the ice. Once on the slippery surface, the deer became easy prey for the ravenous wolves who, with their broader, padded feet, maneuvered with less difficulty.

As a result of the same January flood some of the settlers' cattle met another kind of untimely death. Following an unseasonably warm period, the floodwaters were at their highest when, overnight, it became extremely cold. The cattle that lay in wet and marshy areas froze to the ground before morning. Those standing were frozen up to their knees in the ice. The wolves feasted.[8]

The howling of wolves was an unpleasant sound to early Ohioans. Warren D. Sibley, in his *History of Woodstock, Ohio,* said that in the winter large packs of gray wolves would come out of the dense forest north of Big Darby and raid the plains, serenading the inhabitants with their "demonical howling."[9] Two cousins on a hunting expedition told of setting-up camp for more than a week in an area where the wolves kept them awake night after night. The low mournful wailing, mingled with their howls, continued for hours. Reverend James B. Finley once remarked that one wolf could "make such a chorus of howls as to make you think there are a dozen."[10]

A settler near Dayton who lived in a log cabin with his family told about wolves coming right up to the cabin and scratching at the walls and howling. He said there were many nights when nothing could be heard except the howling of the wolves and the hooting of the owls.

A few pioneer families willingly endured the unharmonious racket for they felt safer at night when they heard wolves howling. They believed it was an indication no Indians were lurking in the forest.

A crew of surveyors blazing a trail through northern Ohio wrote that they found the howling of the wolves "amusing,"[11] but the large majority of people did not share their sentiments and were of the opinion the howling was a hideous cacophony.

It was the adult wolves who did the howling. A pack of howling wolves could be heard from as far away as ten miles. There were two different times of the year when wolves were the most resonant. From late October until the end of February, during the courting season, wolves howled from dusk until dawn. Again in the summer when the pups were out of the den, wolves became equally vocal. Although the pups barked a great deal, full-grown wolves barked only to warn that danger was near. Customarily, they only "woofed."

Wolves, the ancestors of all modern-day breeds of dogs, destroyed poultry, pigs, sheep, calves, and occasionally, foals. During the night, livestock had to be driven into strong enclosures that oftentimes were covered with heavy logs. Even this type of protection for the animals did not guarantee their safety. John Ables of Guernsey County recalled when he was a small child, wolves came in packs and prowled around the cabin, howling relentlessly. One night they broke into a pen which had been constructed against the wall of the house. All their sheep either were killed outright or wounded. As a result of the gruesome sight, he grew up with an overwhelming dread of wolves.[12]

Most families attempted to keep a sheep or two in order to have a supply of wool for weaving but it was almost impossible to raise a whole flock. A few wolves could kill an entire flock in a surprisingly short time. In Medina County, wolves killed over one hundred sheep in one night. Even when deer and other game were plentiful and they were not hungry, wolves killed sheep.

On bright, moonlight nights, people tried shooting at the marauders through the chinks in the walls of their log cabins. The wily animals would approach to within two rods of a cabin, seize a sheep and dash off with it. After the flock had quieted down again, the wolves would return and make another attack.

Mary Guthridge, who came to Ohio with her parents as a young girl, wrote in her memoirs that her father purchased a few head of sheep in 1815. The wolves were so bold, she remembered, the sheep had to be kept inside the cabin at night.[13]

In 1808, James Taylor said that during the night wolves would take vegetables from their garden and washing from off the clothesline. Once they carried away several pewter plates that inadvertently had been left at the spring. Years later, when additional land was cleared for cultivation, the missing plates were found where the wolves had buried them.[14]

Timber wolves, because of their dark, shaggy gray coats, commonly were called "gray wolves." With their powerful jaws they were capable of bringing down prey ten times their own weight and their fangs easily could puncture tough hide. They had a bite pressure of 1,500 pounds per square inch. Usually traveling in packs of two to twenty animals, depending on the availability of game, they had a well-developed plan of attack. Wolves' keen sense of hearing and smell enabled them to detect prey more than five miles away. Some were said to run up to forty miles per hour. Their strength, their agility, and their cunning, in addition to their finely tuned senses, earned the settlers' everlasting hostility.

The manner in which wolves fought often frustrated dogs. Wolves would spring forward, snapping and cutting with their sharp teeth, then retreat, waiting for another opportunity to spring forward again. Their advance and retreat technique was repeated quite rapidly, the wolves snarling all the while. Dogs that were trained specifically in the art of fighting wolves were highly prized.

An 1880 book entitled *How I Know*, written by James Swisher, an Ohioan, included a graphic paragraph about a pack of wolves running-down a deer. He wrote that as he was searching for cattle which had wandered off, he heard the distant howling of wolves. When the wolves came into view he could see, looking down from his vantage point on a hillside, that they were chasing a deer. The deer was outdistancing the wolves but he knew the pack was far too numerous and the deer didn't have a chance. The wolves took turns running and heading off the poor creature and soon began closing in from all sides. Snapping and biting at its slender legs, they succeeded in wounding it sufficiently to bring it to the ground. One of the wolves lunged at its throat and while maintaining the hold, others attacked. Almost in a frenzy they bit into the still-breathing animal. Swisher said that from two hundred yards away he could hear their teeth gnashing together as they devoured it. He counted twenty wolves in the pack.[15]

"Wolves were remarkably daring and impudent," said Reverend James B. Finley. "They would attack grown cattle, and kill colts and two-year-old cows. While hunting the cows one morning in the woods, in company with a lad a little older than myself, we heard a cow bellowing at a piteous rate; and, supposing it was Indians trying to decoy us, we crept up with the tread of a cat, and from the top of the ridge, looked over and saw five wolves hanging onto a large cow, while she was struggling to free herself; and aiming for home. As she came toward us, we took the best position we could and waited their arrival. We fired by concert, killing two and wounding a third. The wolves instantly loosed their hold, ran a few paces and set up a terrible howl. Fearing a fresh supply of this ferocious animal, we ran home, and returning with help, to see what was done, we found two, and tracked the other by its blood some distance. Such were the dangers and hardships to which we were constantly exposed."[16]

Wolves could travel for long periods of time. Their tireless pursuit made a long night's ride on horseback a task to be avoided if possible. As long as the horse moved swiftly, the danger was somewhat minimized. Erastus Stow of Meigs County was not so certain this was true, however, and told about an unforgettable trip to a grist mill. The nearest mill was many miles away and because he walked the entire distance, the round-trip often required an absence of a week or more. Once he had reached the mill, it was not uncommon to wait for several days before it was his turn to have his meal ground. As he was returning home from one of his trips, he realized that he could not hope to reach his house before dark and stopped at an acquaintance's to borrow a horse. At twilight, the wolves commenced howling. Shortly afterwards, he caught glimpses of them slipping through the trees. Ever closer they came until the pack rushed toward them, running and circling around them. The horse, nervously snorting and tossing his head, galloped wildly down the trail. Erastus threw the sack of meal to the ground, lifted his feet as high as he could, grasped the horse about its neck and they fairly flew in the direction of the house, resolutely followed by the wolves. Inside the house, his family heard the commotion and dashed outdoors with fire brands to drive the wolves back into the forest.[17]

Mary Robinson, of Clermont County, started on horseback one cold, winter day for her neighbor's house over twelve miles away. The deep snow slowed her progress to a much greater extent than she had expected and by nightfall she still was in the woods. Snow continued to fall and it soon became so dark she no longer could find her way. When she dismounted in order to more easily find the marks on the trees that guided the way, a pack of wolves began howling and so frightened the horse that he would not stand quietly to let her remount. As the wolves began circling the pair, Mary tied the horse to a tree and to ward them off, waved her arms and walked back and forth all night long, staying far enough from the horse's hooves as he stomped and kicked, and yet near enough to keep the wolves away from her. At dawn, the wolves melted back into the forest and a very weary Mary climbed on the horse and rode to the neighbor's.[18]

A man who lived along Wolf Creek, not far from Marietta, Ohio, frequently fiddled for dances in the community. As he walked home from an engagement one chilly, blustery night in March, he was attacked by wolves. He made a dash for the closest tree which he scaled in record time, his precious instrument tucked safely under an arm. Throughout the long night, shivering and uncomfortable, he sat on a limb high off the ground. Whether he hoped the high pitched tones of the violin might drive the wolves away or it simply was something to while-away the time, "he played the fiddle for them all night until they left in the morning."[19]

In 1818, a bachelor in Erie County borrowed a horse to visit the young woman he was intent on marrying. Traveling back to his own home late at night, he lost sight of the path and as he climbed down to look for the notches on the trees, wolves began howling. The startled horse jerked his head, yanking the reins out of the man's grasp, and sped down the path, leaving the man afoot. The wolves ignored the fast disappearing horse and ran toward the man who sprinted for a tree, climbing as high as he thought he dared. All night he clung to a branch of the tree while the wolves sat below him howling and staring up at him. When the creatures silently fled at first daylight, he was somewhat chagrined to find that the branch he had been clutching so tightly was the upturned root of a gigantic tree which had fallen to the ground.[20]

Enoch Baker, another young bachelor, was returning home late one night after having visited a neighbor's daughter. Leisurely proceeding through the woods, he was jolted out of his reverie when he subconsciously became aware of wolves slipping through the trees toward him. Sensing possible danger, he began running and swinging his club and shouting at the top of his voice. Seemingly unafraid of the commotion he was making, they soon were abreast of him where they alternately fell back and then again raced alongside him, never actually making a concerted effort to attack him. In this extraordinary manner they followed to within sight of the clearing around his home. When they heard the barking dogs, the wolves silently disappeared.[21]

James Gilliland, of Van Wert County, experienced a different sort of situation when he lost his trail in the darkness. Unable to see, he lay down to wait for the moon to rise. Although very tired, he intended to rest only until it was light enough to see again. A few yawns later, however, he was sleeping soundly when, all of a sudden, he felt a cold nose pressed against his face. In a flash, he jumped to his feet. He could distinguish the dim outline of an animal moving away from him but it was not until it stopped and began to howl that he knew it was a wolf. By now the moon was casting a yellowish glow through the treetops and, thoroughly shaken, he hurriedly made his way home.[22]

Even when the moon shone brightly elsewhere, in dense forests where trees grew closely together, the light was obscured and the forest floor would be in shadows. So it was when James Criswell, a blacksmith by trade, encountered wolves. He saw what he supposed in the uncertain light were tree stumps but when they began moving—and moving toward him—he realized they were wolves. As he succinctly expressed it in later years, "He then struck out with all possible speed for his cabin, pursued by a large pack of wolves."[23]

Samuel Huntington, who was governor of Ohio in 1809 and 1810, had an unexpected skirmish with wolves in 1813 while riding on horseback to Cleveland. Although Cleveland already was the nucleus of an enterprising commercial center, much of the town was covered with trees and the huge chestnuts that skirted today's Superior Street were so thick they completely blocked Lake Erie from view. Near the town was a swamp, a favorite haunt of wolves and it was here the ex-governor had his confrontation with them. Carrying only a tightly furled umbrella and his leather case filled with legal documents, he was attacked by a pack of wolves. With no weapon to protect himself, Huntington used the umbrella as a club and over and over again struck at the snarling animals. Quite by accident, the umbrella suddenly flew open, startling the wolves to such a degree they fell back in alarm and dashed off into the trees. Gathering up what was left of the umbrella, Huntington rode on into town.[24]

One account regarding Governor Huntington states that the attack occurred in the daytime. His terror-stricken horse, it was said, came to a dead stop and, trembling violently, refused to move. Following the attack, the governor calmed the still quivering horse and then rode on into town.[25] Another report says the attack happened after dark and the horse was so frightened it "outstripped the wolves and bore his rider off in safety. Huntington said that due to the fleetness of his horse, he owed the preservation of his life."[26]

Traveling through a thick woods during the daytime could be a risky undertaking in many places in Ohio. Thomas West was a schoolboy in Harrison County who nearly every day walked three miles to and from school. The path he used crossed a woods where wolves often were a threat and a small boy traveling alone could be in eminent danger. Although the trees were clearly marked, the light was dim in the thick stand of timber. When it was overcast or foggy, it was safer for Thomas to remain at home and not try to cross the woods. Because it was mandatory he return home before dusk in order to avoid the wolves, he always left school even on the sunniest days before afternoon classes began.[27]

Levi Rowland had a recurring and disturbing dream during the night about fighting a wolf. The morning following his dream, he dressed hurriedly and left the house early to find some of his cattle that were missing. He could hear them lowing somewhere ahead of him when he remembered his dream of the night before and, feeling just a bit sheepish, he nonetheless retraced his steps to a grove of trees and armed himself with a heavy stick. Turning again in the direction the cattle had gone, he took but a few steps when a wolf sprang from behind a thorn bush and, barring its teeth, lunged at him. Striking it repeatedly with his club, Levi finally knocked it to the ground and beat it to death. Now thoroughly winded, he took a deep breath, wiped the perspiration from his forehead, gripped the stick, and resumed his search for the cattle he could hear in the distance.[28]

In Ashtabula County, Elijah Thompson was hopeful of shooting a bear when his dog followed a scent deep into the forest. Although the dog soon was out of sight, Elijah could hear his excited barking. When the barking abruptly changed to yelps and then to whimpers, he knew something was very wrong. He quickened his steps and soon saw that seven full-grown wolves had surrounded the dog and already had inflicted serious wounds. For fear of hitting the dog, he pointed his gun in the air and fired, fully expecting to frighten the wolves. The loud blast, however, had little effect on them. Taking no time to reload, he gripped the gun by the barrel and charged into their midst, striking right and left until they backed off and fled. Cradling the badly injured dog in his arms and carrying the broken rifle over his shoulder, Elijah made his way home.[29]

George Warth and Peter Niswonger were looking for game along the Shade River in Meigs County when they decided to hunt in different directions. George Warth began walking along the side of a ridge and soon noticed a pack of wolves near the river in front of him. He was aware that they were watching him closely but because they appeared more curious than threatening, he did not feel the situation was a dangerous one. When the largest and most aggressive wolf began to lope toward him, followed more slowly by the others, however, he unhesitatingly shot the leader. Instantly, the rest of the pack rushed at him. Surprised, for he had assumed the wolves would run away, George quickly waded into the river, at the same time reloading his rifle. Turning around, he was surprised again to see that they actually were swimming toward him. He shot a second wolf. Still, the others swam toward him. There now was no time to reload and he was hip deep and beginning to flounder in the water. He was alert to his very real danger—not only from the wolves but also from the risk of drowning. He frantically beat them off with the stock of his gun which almost immediately broke. Using the barrel as a club, he finally drove the remaining wolves back to shore where they sat on their haunches, watching his every move. George had had enough. He struggled across the river to the opposite side where he sat and waited for Peter. Reluctant to face the wolves again, and with only one useable gun between them, "the hunters returned to their homes on Oldtown creek. Next day they increased their force and went back to the place of the battle and found two dead wolves but no live ones."[30]

Abraham Studebaker of Darke County had butchered a hog. Sometime in the night he heard fighting among wolves that had been attracted by the smell of the fresh blood and were congregating outside his cabin. There was a sudden, agonized yelp followed by vicious snarling and terrible growling. Then, just as abruptly, there was total silence. Although puzzled, no one ventured outside to investigate until after daybreak. In the morning, they found a dead wolf lying beside the door, its tongue frozen to the blade of the axe that had been used to slaughter the hog. "Bits of flesh and blood still clung to the blade. The other wolves had killed the wolf."[31]

Jacob Simpson, a young boy in Berlin township in Erie County, was alone one evening and for lack of something better to do, sat on a log near the house petting his dog and playfully howling like a wolf. The dog pricked up its ears and also began to howl. Laughing and in high spirits, the boy continued his duet with the dog until in the deepening twilight answering howls from real wolves sounded disturbingly close. Jacob made a dash for the door, whistling for the dog to come along. The dog, however, remained where it was and watched the approaching wolves with interest until several of them darted toward it. Then it turned and, with tail tucked between its legs, ran for the door which Jacob was holding open. It flew inside followed so closely by the wolves they nearly slipped through the door at the same time. Listening to the now-not-so-amusing dissonant chorus outdoors, the boy barricaded the door with a pile of firewood as the howling wolves circled round and round the house.[32]

Leonard Bousman and his sons worked until after nightfall clearing away fallen branches from under the trees on land where they planned to cultivate a field of corn. To ward off any predators that might be hidden in the shadows, they piled the treetops and broken limbs in huge piles and burned them. In the flickering light of the fires, not long afterwards, the men noticed wolves at the edge of the forest but they felt reasonably safe. One of the wolves, against all expectation, gave a tremendous leap and sprang at the older man who struck it with his ax. The boys rushed to their father's aid and quickly killed the animal.[33]

Ephraim Magill had earned the reputation not only as an accomplished wolf hunter but also as a spell-binding raconteur. Although the stories ordinarily were about his own successful exploits, one of his favorites was about a misadventure. After locating a wolf den in a hollow tree and observing an adult female leaving it, he decided to see if there were wolf pups inside and, if possible, take them home for pets. The entrance to the den was an opening several feet above the ground and as he began to peer inside, he became apprehensive for some reason and, withdrawing his head, cautiously poked his walking stick inside. Instantly, something savagely bit off the end. Ephraim did not tarry but turned and walked away. At the conclusion of the tale,

Ephraim would laugh uproariously and launch into another story.[34]

Any number of McGill's experiences turned into risky ventures. He told about once observing a mother wolf standing beside a hollow log where her pups were hidden. He shot the wolf and on hands and knees, tightly gripping his knife, started to crawl inside the log to find the pups. Suddenly, a second female wolf began to rush past him. He stabbed it with his knife and, pulling it out of the way, crawled further into the hollow log where he found and killed five pups. For the scalps of the five pups and the two full-grown wolves he received twenty-eight dollars in bounty money.[35]

Ephraim Magill eventually did acquire a wolf pup which he raised and trained to howl on command. In the spring and early summer he would tramp through the woods with his pet searching for wolf dens. Finding a promising location, he would wait until the time of day when he knew a mother wolf ordinarily returned to the den to feed her young. At the signal, his pet began to howl and, invariably, the she-wolf answered. Magill could pinpoint the site of the den and the following morning would kill the female wolf and capture the pups.[36]

In 1882, a physician in Williams County received a tiny wolf pup as a gift. It grew to become a friendly and endearing pet who was especially fond of the doctor's small daughter. The two played together hour after hour. They were frolicking one day when the wolf playfully grabbed the child's doll and hid it under a table. The child began to cry; whereupon, the wolf carried the doll back and laid it at her feet. "Then when she took it up again, he jumped and capered around her as though he could scarcely contain himself for joy." As Ephraim Magill had done, the doctor taught his pet to howl upon command. "It was a most horrid noise, which became at last such a nuisance to ourselves and neighbors that we were obligated to get rid of him."[37]

John James Audubon, the noted artist in the early 19th century, was one of the first to study and to paint native American birds in their natural habitat. Once while sketching in the woods, he is said to have met a man with a tame wolf trotting at his side. When Audubon expressed great surprise at seeing the creature, the owner claimed the wolf was a well-behaved animal who trailed deer far more satisfactorily than any dog he'd ever owned.

John Chapman, more familiarly known throughout Ohio as Johnny Appleseed, was the inspiration for many animal-related stories. Some of the incidents actually may have occurred; others doubtlessly did not. One of his biographies stated that he once found a wolf caught in a trap and after releasing it, tended to its injury and nursed it back to health. It became a pet and followed him about like a dog until someone shot it by mistake.[38]

Understandably, however, relatively few Ohio pioneers felt kindly enough toward wolves to keep them as pets. To the contrary, wolf hunts became popular pastimes.

In 1819, settlers in the area of today's Mechanicsburg, Ohio, gathered for a wolf hunt. Many of them were losing livestock and they believed much of it was due to wolves. On a specified day, the men positioned themselves in a ten-mile-wide-radius and simultaneously converged toward the center. A few of the men were selected to carry guns but the majority were armed only with clubs and pitchforks. At intervals, horns were blown to indicate the location of the participants. When, after several hours, they approached the center, they found not a single wolf.[39]

A more successful wolf hunt was held near West Jefferson. Several wolves were shot and killed; however, there also was an accident. Reverend Isaac Jones, a minister who was asked to join the hunt because of his skill with a rifle, was appointed to be one of the marksmen. As the hunters drew near, he climbed a tree in order to have a clearer view.

While his rifle was being lifted up to him, the hammer caught on someone's coat sleeve and the gun discharged, striking the Reverend in the hand and permanently crippling it.[40]

Possibly the best known game hunt in the State of Ohio took place in Medina County on December 24, 1818. Soon after sunrise a signal was given and, with guns ready, over six hundred men and boys who had assembled, some from considerable distances, began advancing toward a "half mile limit." Although some animals escaped, approximately one hundred were killed before reaching the "half mile limit." "The half mile limit" was a perimeter marked by blazed trees beyond which guns could not be fired for fear of shooting persons approaching from the opposite direction. Inside the cordoned area, dogs were released to flush the remaining game. A tally at the end of the hunt enumerated seventeen wolves killed, twenty-one bears, and three hundred deer. The number of turkeys killed was not taken into account. Amazingly, the only accident was a flesh wound from a load of buckshot which ricocheted.[41]

Medina County Wolf Hunt - December 24, 1818
Henry Howe, *Historical Collections of Ohio*

Throughout the state, wolves, regardless of age, sex or size, were killed in great numbers. In the community around Mt. Vernon wolves were so troublesome in 1805 that the men signed an agreement and together pledged nine bushels of corn for every wolf scalp collected. The first winter, forty-one wolves were killed.[42]

Male wolves, some of whose paw prints could be as large as an adult man's hand, frequently weighed well over one hundred pounds. Females weighed appreciably less. The female wolf that Abraham Inlow saw crossing an open prairie in Clarke County was a small one and, on horseback, he easily overtook her. When she attempted to hide in a clump of tall grass, he dismounted, grasped her by the hind legs, and pulled her toward a stand of timber. Struggling unsuccessfully to twist her body into a position to bite him, she growled and snarled in rage. Inlow dragged her to a large walnut tree where he lifted and, with all his might, flung her with such force against the trunk of the tree, it smashed her skull.[43]

Wolves were a trial to pioneers throughout the entire State of Ohio and most counties offered a bounty of $2.00 to $6.00 for each wolf scalp. Brown County required every hunter to produce the wolf's head, with ears intact, before a person could claim his money. In 1799, it was mandatory for hunters to take an oath: "I do solemnly swear (or affirm) that the head now produced by me is the head of a wild wolf, taken and killed by me in the county of _____, within six miles of some one of the settlements within the same to the best of my knowledge, and that I have not wittingly or willingly spared the life of any bitch wolf, in my power to kill, with the design of increasing the breed. So help me God."[44]

In 1809, certificates for wolf scalps were issued which could be used as currency and also were accepted as payment for taxes. This proved to be an even greater incentive for hunters to kill as many wolves as possible.

When sixteen year old William Wolvertin of Preble County shot and killed a wolf, he used the $4.00 bounty money to purchase a coat, the first "store-bought" coat he ever owned.[45]

The bounty system in Champaign County was so effective that Colonel John H. James, an attorney and politician from Urbana, wrote, "The fund for wolf and panther scalps equaled the amount given to all the associate judges in a year's time."[46]

Because wolf skins served no useful purpose except, it was said, to make drumheads, they were of no monetary value. Wolf scalps, however, continued to be of financial significance. One pioneer who knew the site of a wolf's den, would kill the pups year after year when they were nearly grown and collect the bounty money. Eventually, someone else discovered the den and killed not only the pups but also the parent wolves.[47]

Tom Rogers was a man who relied on bounty money. He specialized in eliminating wolves and for over twenty years was known far and wide. "Old Tom" was not an easy man to forget. He stood six feet tall and had long black hair and a long black beard. He wore a wolf-skin cap with the tail hanging down the back, buckskin breeches, and moccasins. From his wide leather belt hung a large knife and an always-sharp tomahawk. He carried a heavy rifle, a pouch, and a powder horn. He lived for months at a time in the forest, often staying at night in a sycamore tree. Although he remained relatively close to settlements and frequently exchanged venison and wild turkey for bread at a nearby cabin, he spent the days running his trap lines and tending to his deadfalls and his wolf traps. He generally traveled to town twice a year to collect the bounty for his wolf scalps and to sell his mink, otter, and raccoon furs. He was known to purchase large quantities of asafoetida, a fetid resin he used to coat his traps, for the putrid odor attracted wolves.[48]

Although different people favored different methods of capturing wolves, setting steel traps was one of the most common. They could be found at nearly any trading post or general store. They were not infallible, however, for wolves occasionally gnawed off a leg in order to escape.

Hunters were well aware of wolves' highly sensitive olfactory capabilities and tried in various way to disguise their own scent when setting traps. Some tried leaning over from astride a horse and wore gloves soaked in animal blood. Others boiled their steel traps in oak leaves and when setting them, stood on pieces of leather.

Wooden wolf traps, usually referred to as "wolf pens," easily could be constructed by the settlers themselves. The pens were described as log structures, six feet long, four feet wide, and three feet high with heavy wooden floors and lids. Meat, almost any kind of meat except that of another wolf, was placed inside as bait and the lid propped open. By means of a spring, the lid would fall when the wolf crept inside, thereby imprisoning it.

An elderly gentleman in Delaware County once crawled into his wolf pen to make an adjustment and it sprang shut, pitching him face down onto the floor. Unable to lift the lid, he lay in his uncomfortable position all that day and through the following night until a passing hunter heard his moans and released him.[49]

Deep pits, called deadfalls, sometimes were dug and camouflaged. For several years a wolf in Madison County eluded residents as it raided sheep in the area. Finally, the den was located and adjacent to it a pit was dug. It was covered with a board, arranged on a pivot, and bait was suspended from a tree branch over the pit. The ruse succeeded and the wolf was captured. There was, however, an unexpected consequence for after the wolf was caught, the pit remained open and a cow stumbled into it. Many months later and after a long and fruitless search for the cow had been made, her skeleton was found. Only then was the pit filled-in.

There was great rejoicing when the Madison County wolf was captured. It was not killed but was removed, bound and muzzled, and carried to a nearby grist mill where it was chained and kept in the courtyard. Customers at the mill were encouraged to turn their dogs loose and watch as they fought with the wolf.[50]

Snares occasionally were used to trap wolves. The top branches and foliage of a sapling were cut off and the upper trunk secured with a strong cord which was adjusted to form an open slip-noose on the ground. The sapling then was bent nearly double and fastened so that the slightest contact would pull the noose tight, theoretically catching the wolf by the neck. Wolves quickly became suspicious, however, and instinctively avoided them.

The pioneers with their deadly rifles were responsible for the demise of untold numbers of wolves in Ohio. Men nearly always kept their guns close at hand—when plowing or hoeing in their corn patches, when cutting wood in the forest, and, of course, when hunting for game. It was a natural inclination to shoot at every wolf they saw. Powder and lead were expensive items to purchase but in order to survive in the forest, guns were essential. One man lamented that he had paid more for his guns and ammunition than he had for the combined total of his mules, wagons, and provisions. Young boys learned to shoot as soon as they were large enough to handle the heavy weapons and it was not unusual for them to be punished if they missed the target.

Guns, on the other hand, sometimes assisted wolves in their quest for food. Reverend David Zeisberger noted that Indian hunters ordinarily killed fifty to one hundred and fifty deer each fall because there was a profitable trade in deer hides. When deer were plentiful, an Indian ate only the most desirable parts, skinning the deer on the spot and abandoning the remaining meat for the wild animals to consume. Wolves, adaptable and opportunistic creatures that they were, learned to follow the hunters and move in the direction of the shooting where they, in all probability, could partake of a readily available meal.[51]

All along the Ohio frontier wolves were hated. The first people to come to the territory were interested primarily in trapping and hunting. Soon afterwards came immigrants who were more concerned with clearing and establishing homes in the wilderness. Both goals represented the dreams of many and wolves had no place in either of those dreams. Wolves were considered an abomination to be eliminated as quickly as possible. They were tracked down with such relentless determination that, cunning as they were, these masterful hunters were outnumbered and eventually disappeared. Hundreds were killed. Others moved ever westward and northward in search of wooded regions devoid of mankind.

Our forefathers would have been in complete agreement with the droll assertion printed in an Ohio history that "Wolves, rattlesnakes, and wildcats were uncomfortably numerous."[52]

Chapter 2: PIONEERS HATED WOLVES

1. W. H. Beers., *History of Champaign County* (Chicago: W. H. Beers & Co., 1881), p 330.
2. William Dean Howells, "The Captivity of Boone and Kenton," *Stories of Ohio* (Cincinnati, Ohio: American Book Co.,1897), p 58.
3. Henry Howe, *Historical Collections of Ohio*, vol. 2 (Cincinnati, Ohio: Krehbiel & Co,1904), pp 271-273.
4. ——, p 503-504.
5. Henry Clay Alder, *A History of Jonathan Alder: His Captivity and Life With The Indians*, 1870; ed. Larry L. Nelson (Akron, Ohio: University of Akron Press,1999), p 74.
6. ——, p 74.
7. Johnda Davis, *Journal of Jonathan Alder* (Columbus, Ohio: Ohio Historical Society, 1988), p 97.
8. Henry Howe, *Historical Collections of Ohio*, vol. 2 (Cincinnati, Ohio: Krehbiel & Co.,1904), p 469.
9. Warren D. Sibley, *History of Woodstock, Ohio* (n. p., 1907),p 11.
10. James B. Finley, *Autobiography of Reverend James B. Finley* (Cincinnati, Ohio: E. Thompson Printer, 1853), p 93.
11. Henry Howe, *Historical Collections of Ohio*, vol. 2 (Cincinnati, Ohio: Derby, Bradley & Co., 1848), p 345.
12. ——, vol. 1(., 1896), p 733.
13. Warren Everhart, "History of Mary Guthridge" (n. p., 1956).
14. John H. James, "Reminiscences of James Taylor," John H. James Collection (n. p.).
15. James Swisher, *How I Know* (Cincinnati, Ohio: Press of Jones Bros., 1880),pp 230- 231.
16. James B. Finley, *Autobiography of Rev. James B. Finley* (Cincinnati, Ohio: E. Thompson Printer, 1853), p 38, 39.
17. Stillman C. Larkin, *Pioneer History of Meigs County* (Columbus, Ohio: Berlin Printing Co,1908), pp 81- 82.
18. Henry Howe, *Historical Collections of Ohio*, vol. 1(Norwalk, Ohio: Laning Printing Co.,1896), p 410.
19. S. P Hildreth, *Biographical and Historical Memoirs of Early Pioneer Settlers of Ohio With Narratives of Incidents and Occurrences in 1775* (Cincinnati, Ohio: Derby & Co, 1852), p 446.
20. Lewis Cass Aldrich, *History of Erie County* (Syracuse, N.Y: Mason & Co,1889), p 516.
21. William Dean Howells, *Stories of Ohio* (Cincinnati, Ohio: American Book Co. 1897), p 167.

22. Thaddeus Gilliland, *History of Van Wert County* (Chicago: Richard & Arnold, 1906), p 147.
23. W. H. Beers, *History of Madison County* (Chicago: W. H. Beers & Co., 1883), p 632.
24. Joshua Antrim, *History of Champaign and Logan Counties, From Their First Settlement* (Bellefontaine, Ohio: Press Printing Co. 1872), p 24.
25. ——.
26. Henry Howe, *Historical Collections of Ohio* (Cincinnati, Ohio: Derby, Bradley & Co, 1848), p 280.
27. ——, vol. 1 (Norwalk, Ohio: Laning Printing Co.,1896), p 552.
28. Thaddeus Gilliland, *History of Van Wert County* (Chicago: Richland & Arnold, 1906), p 137.
29. Henry Howe, *Historical Collections of Ohio* (Cincinnati, Ohio: Derby, Bradley & Co., 1848), p 47.
30. Stillman C. Larkin, *Pioneer History of Meigs County* (Columbus, Ohio: Berlin Printing Co., 1908), p 47.
31. W. H. Beers, *History of Darke County* (Chicago: W. H. Beers & Co.,1880), p 417.
32. Lewis Cass Aldrich, *History of Erie County* (Syracuse, N.Y.: Mason & Co., 1889), p 441.
33. W. H Beers, *History of Miami County* (Chicago: W. H. Beers & Co., 1880), p 405.
34. Thaddeus Gilliland, *History of Van Wert County* (Chicago: Richland & Arnold,1906), p 168.
35. ——.
36. ——.
37. Henry Howe, *Historical Collections of Ohio,* vol. 2 (Cincinnati, Ohio: Krehbiel & Co.,1904), p 856.
38. Robert Price, *Johnny Appleseed: Man and Myth* (Urbana, Ohio: Urbana University,1954), p 171.
39. W. H. Beers, *History of Champaign County* (Chicago: W. H. Beers & Co., 1881), p 595.
40. ——, *History of Madison County* (Chicago: W. H. Beers & Co., 1883), p 352.
41. Henry Howe, *Historical Collections of Ohio,* vol. 2 (Cincinnati, Ohio: Krehbiel & Co, 1904), pp 203-206.
42. ——, vol. 1 (Norwalk, Ohio: Laning Printing Co.,1896), p 983.
43. W. H Beers, *History of Clark County* (Chicago: W. H. Beers & Co., 1881), p 739.
44. ——, *History of Brown Count* (Chicago: W. H. Beers & Co., 1883), p 269.

45. *History of Preble County* (Cleveland, Ohio: H. Z. Williams & Bros., 1881),p 249.
46. Newspaper clipping (n. p., n. d.).
47. *Ohio Historical Review*, vol. 9, no 42 (Columbus, Ohio: pub/ed, Ralph S. Estes , n. d.).
48. W. H. Beers, *History of Miami County* (Chicago: W. H. Beers & Co., 1880), pp 239, 388.
49. Henry Howe, *Historical Collections of Ohio* (Cincinnati, Ohio: Derby, Bradley & Co, 1848), p 275.
50. W. H. Beers, *History of Madison County* (Chicago: W. H. Beers & Co., 1883), p 633.
51. Rev. David Zeisberger, *David Zeisberger's History of the Northern American Indians,* ed. Archer Hulbert & William Schwarze (1910; reprint, Lewisburg, PA: Wennawoods Publishing, 1999), p 14.
52. W. H. Beers, *History of Montgomery County* (Chicago: W. H. Beers & Co.,1882), p 290.

"PAINTERS" AND WILDCATS

Chapter 3

Early Ohio settlers sometimes feared panthers more than wolves. These bold, defiant creatures could be found lurking everywhere and because of their stealthy nature often escaped detection. They prowled roads and invaded farmyards. Stretched out on overhanging, low-lying branches of trees or concealed in the underbrush, they took their prey by surprise.

Panthers had a head and face like that of a domestic cat, a long tail, and paws that were armed with sharp-edged claws. Their eyes were large and the pupils of corresponding size which permitted clear, nocturnal vision. They had an extraordinary hearing range, far more acute than that of human beings. Their judgment when leaping was highly accurate and once confident of capturing their quarry in a single leap, would spring, grip their victim with their claws, and throw it to the ground. They used their long dagger-like incisor teeth for stabbing. The wide gap between the incisor teeth aided in seizing and securing their intended meal. They had tremendous bite force. Attempting to sever the jugular vein, the big felines seldom released their hold until their prey was dead.

Panthers were powerful killing machines. David Zeisberger wrote that they were "beasts of uncommon strength."[1] They were strong enough to carry a live hog up into a tree or through deep snow. Except for the jaguar, panthers were the largest of the American cats. Although buffalo had few natural enemies, other than man, panthers were known to attack them, dropping from a tree onto their backs and clutching the skin through the huge animals' thick, shaggy coats. Daniel Boone told of once watching a buffalo try to shake a panther from its bleeding back. He shot the cat.

Pioneers along the Ohio frontier knew panthers variously as pumas, catamounts, mountain lions, cougars, and the more colloquial term, "painters." They were hunted because they were dangerous predators, not for the meat they could provide. Although nearly everyone considered their flesh unacceptable, there were those who thought it had a sweet taste and did not object to the meat. The general consensus of opinion, nevertheless, remained one of aversion. When a group of prospective land purchasers lodged for the night in a tavern along the Muskingum River, they learned after the meal was over that the disagreeable meat they had been served was panther. They were not pleased and loudly voiced their displeasure.

Bounties, i.e., rewards paid in order to encourage the elimination of predatory animals, were offered for panther as well as for wolf scalps. Panthers could be difficult animals to kill, however, as John Sloan knew only too well. He and his dogs were trailing a deer along the fringes of the woods when the dogs treed a panther. While it was distracted by the excitedly barking dogs, Sloan shot it, injuring but not killing it. Still, the cat sprang at him. The dogs leaped up and intercepted the spring and although it was mortally wounded, the snarling panther fought them ferociously. Sloan later marveled that it had required nine, well-placed shots to finally kill the beast.[2]

In 1821, Thomas Cartmill was concentrating on a flock of wild turkeys as they began perching in a tree for the night. He heard a slight noise in back of him and whirling, saw a panther crouched and preparing to pounce upon him. He swiftly leveled his gun and, as it sprang, shot it through the heart. Its leap was of such tremendous force that in its death grip, it clung for several minutes to the tree in front of which the man had been standing.[3]

A Darke County pioneer had a comparable experience. He was working near the edge of the clearing around his house when he glanced upward and saw a panther on the limb of a tree with muscles tensed, ready to spring at him. He raced for the gun he had leaned against a tree and as the animal hurled toward him, he fired. The bullet struck it in the shoulder and as it struggled, teeth bared threateningly and spitting and snarling with fury, he reloaded his rifle, took careful aim and killed it.[4]

A man who homesteaded in Montgomery County at the beginning of the 19th century heard prolonged squealing and hurriedly went to check on his hogs . He could find nothing that would have caused the commotion but as he stood watching them, something fell from a tree. There at his feet lay one of his hogs. On the branch above was a panther glaring at him in unblinking intensity. The man was not carrying a gun but instinctively knew that he dare not turn his back on it to try to run away. Never flinching, he stared back at the crouching feline. Finally, little by little, it descended the tree and veering aside, slipped into the woods.[5]

David Zeisberger, the dedicated missionary who worked most of his life with the Delaware Indians in Ohio, wrote a book called *History of the Northern American Indians* and in it described the wildlife that was native to the state. He said, "One must not take his eyes off a panther, and if he has not the courage to shoot, gently walk backward, until he is a good distance away. If he shoots and misses, then he is in imminent danger and must keep his eyes fixed on the animal."[6]

Lucas Sullivant, the man who founded Franklinton in 1797 on the west bank of the Scioto River, also was a widely-known surveyor and did extensive surveying in the State of Ohio, especially in the Virginia Military District. (Franklinton comprises a portion of the west side of present-day Columbus,

Ohio.) Sullivant is reported to have said his most hair-raising incident with a wild beast may have been one that took place while he and his crew were camping for the night at a salt lick. They were relaxing around a fire when one of the men noticed a tail swinging back and forth directly overhead. There on a tree branch, with ears flattened, squatted a panther malevolently regarding them. The man reached for his rifle, "aimed at the head, between the glaring eye-balls of the animal, and, with a steady hand, pulled the trigger." Sullivant noted in his journal that the panther fell into the fire and in its death throes, scattered embers and glowing pieces of wood.[7]

A surveying crew in what was to become Madison County had a scare with a panther in 1795. Only four days previously, this same group had a harrowing experience with an Indian scouting party and barely had escaped with their lives.

Jonathan Alder, early one summer while living with the Indians, was attempting to imitate the bleating of a fawn in order to draw a doe within shooting distance when he noticed movement in a bush." It would stop and then I would blate again, and it would start and come a little further and stop. I would blate again and so on until it got quite close. I squatted down behind a large bush so as not to be seen until I could call it up close to me. It seemed to come very cautiously, more than common. I would blate a little every now and then, and it came right to the opposite side of the same bush that I was behind. I could see the brush shaking, but the deer didn't appear to be moving. The brush was so thick that I couldn't see it, so I gradually raised up and, to my great horror, there was a huge panther, his tail shipping from side to side. This was what had kept the brush shaking. His feet were all on the motion, just as you have when a cat is about to pounce onto a bird. In less than a minute, if I hadn't risen up, he would have been on me. I was so scared that I never thought of shooting, but hallooed with all my might. I guess the panther was about as bad scared as I was for his first leap must have gone twenty feet. He went off in a hurry and I left the chase for that day. There was no more deer blating by me that summer."[8]

It was standard practice for men in a large hunting party to separate and several of them go together in different directions, meeting from time to time at a base camp. The leader of one such a party who happened to be working alone at the moment was awakened during the night by the loud whinnying of the packhorses. Sensing they might be in danger, he jumped up and, quickly lighting a torch at the fire, took his gun and ammunition and raced in the direction of the horses. He could hear them snorting and thrashing about among the trees where they were corralled.

Then, he saw it — the shadowy figure of a panther inching forward with body held low and intently watching the horses. The panic-stricken horses continued to mill back and forth in the line of fire but, holding the torch as steady as possible, he at last seized an opportunity and pulled the trigger. The bullet hit the cat and from the impact, the big animal was tossed several feet into the air. It landed on all four feet and sank down to the ground, lashing its tail in fury and screeching. The second shot was a misfire. Dropping the gun, the man threw his tomahawk with a powerful thrust. It struck deep into the snarling beast's skull but still the animal spat and prepared to spring at him. Growing increasingly concerned, he unsheathed his knife with the intention of stabbing it when, all at once, it gave a long shudder and slumped motionless to the ground. Later in the season when everyone was assembled at the base camp, the story about the panther was told and retold.[9]

Relatively young boys often were adept at handling firearms. John Williams, a thirteen year old, shot at a seven foot panther near his house late one afternoon. Although it soon eluded him among the thick undergrowth, he managed to shoot it one more time. From the amount of blood, it was obvious the animal had been wounded but by now it was far too dark to continue the chase and, reluctantly, John headed for home. Early next morning, he bounded out of bed eager to pick up the trail and went into the woods where he followed the injured cat's path of blood to a hollow log. Prudently shooting it again, he began chopping open the log. Astonished to find it still alive, he shot it a fourth time! Reassuring himself it now was quite dead, he dragged it home "where they found the previous day's bullet had pierced the panther's lung. It measured seven feet from tip to tip."[10]

According to a history of Clermont County, Adam Bricker of Williamsburg, Ohio, killed an eight foot panther, the last panther killed in the region. While undertaking to decoy a doe by imitating the cries of a fawn, "he was confronted with a ferocious panther. It required quick work to save himself from being torn to pieces. Fortunately, at his first fire the panther fell dead."[11]

In 1805, a panther was killed on what today is Euclid Ave. in Cleveland. From its nose to the tip of its tail it measured nine feet. Another nine foot panther was killed some time prior to 1815 on the road between Urbana and Springfield.

Opposite the modern-day city of Ashland, an eleven foot panther was killed. In the spring of 1803, the Jonathan Melvin family was traveling by pirogue, a type of long dugout canoe, on the Ohio River. Late in the afternoon they went ashore to make a temporary three-sided camp for the night. In front of the open, fourth side, they built their campfire. After preparing and eating a simple meal, they wrapped in blankets and bedded down on the ground inside the structure. They placed the baby beside her mother closest to the warmth of the fire. Father and son slept on the other side of them. Their dog, a huge, white mastiff, lay nearby.

When the fire had burned low and only a soft light was glowing faintly through the embers, a panther silently slipped past and sank its teeth into the sleeping baby. The mother awoke and screamed, a startled, piercing cry. The dog jumped up and leaped at the cat. The panther dropped the baby and spun around to slash at the dog. The father lit a pine knot and thrust it into the cat's face; whereupon, the cat, flattening its ears and abandoning the dog, jumped up a tree nearby. Mr. Melvin then threw a fire brand into the tree, driving the animal up a bank and onto a rock ledge. Its snarls reverberated into the night, convincing the parents there must be other panthers lurking in the dark. Afraid to remain where they were, they reloaded the boat and poled on up river.

The baby's injuries, although serious, were not life-threatening. The panther's lower teeth had ripped open the flesh on the infant's right shoulder and the curving upper teeth had lacerated the scalp in several places but, amazingly, had done little other damage.

The following morning the family sought help from people they saw at a cabin in a small clearing. A number of the men traveled to the campsite of the night before where they discovered the beast who "jumped up a tree at the foot of the hill, climbed to a big limb about thirty feet from the ground, and perched there with his rump against the tree and his eyes on the men." At a signal, two of the men fired together and the panther fell, "both bullets going into the panther's head. He fell among the dogs and lacerated both of them before his muscles relaxed and he died."[12]

William Robinson of Fayette County claimed to have killed an eleven foot panther. While hunting not far from his cabin, he realized a panther was following him and, jumping behind a tree, cocked his rifle, leaned out, aimed, and fired. Although he knew he had hit it, the big cat barely slackened speed but hurtled toward him. He fired a second time and struck it between the eyes. It fell with a thud.[13]

David N. Morris, Sr., of Miami County, was said to have killed three panthers in one morning. Morris, known to be an avid hunter, was determined his sons be as expert as he and sometimes boxed the boys' ears when they did not perform to his satisfaction.[14] Boys as young as ten often were expected to carry arms.

During the latter half of the 1850s, pioneers in Clark, Miami, and Champaign counties met together and held annual "pioneer meetings." At an 1878 meeting in Addison, modern-day Christiansburg, Ohio, an often-told story was repeated about a woman who, many years before, had been "chased for over a mile, while on horseback, by a panther, which tore all her clothing off: saved her babe by holding it high in her arms."[15]

Rarely did one hear of panther cubs being captured but a man in Washington County did just that. He was one of several hunters employed by a surveying team to supply fresh meat for their meals. The account in Howe's *Historical Collections of Ohio* offered no details, saying only that one day a hunter "looked into a cave and found two panther cubs. He put them in a bag and afterward exhibited them in New Orleans."[16]

Those who heard panthers scream compared the sound to the most horrifying shrieks one could imagine. It was unnerving for even the most stout-hearted. For children, it was particularly terrifying. Margaret Ward Banes, the oldest daughter of Urbana's founder, William Ward, Sr., said that she and a younger sister "rambled in the wild forest and were badly frightened by the cries of a panther."[17] All that remains of the "wild forest" about which she spoke is today a grove of trees in the vicinity of Urbana University.

Warren Everhart, one-time Champaign County recorder, periodically contributed articles for a local newspaper. He wrote about early residents, places, and events. A column about his grandparents told the story of a pair of panthers in 1805 that prowled outside their home near Catawba. The hideous screams were "chilling," they remembered.[18]

Ray Crain, noted Ohio historian and author, once recounted his great-grandmother Crain's haunting recollection of a panther. The family had finished butchering a hog and during the evening a panther, possibly scenting the fresh meat, tried to force its way into the cabin. Clawing at the door and intermittently yowling and snarling, it would shove its front feet under the door. Fearing the determined and powerful creature might succeed in forcing its way inside, the unflinching pioneer woman grabbed an ax and chopped off one of its paws.[19]

****

Although our progenitors wrote far less about their encounters with wildcats, there were a few incidents. Often called "bobcats" because of their short, stubby tails, these carnivores usually ranged in length from two and a half to five feet and stood about fifteen inches high at the shoulders. Their odious squalling was unbelievably frightening. They moved like shadows, pausing frequently, and their coats blended so well with the surrounding foliage they often were not noticed. They were silent and highly efficient stalkers but, unlike other predatory mammals who depended more on the sense of smell to detect prey, wildcats relied to a greater extent on their eyesight and on their hearing, both of which were exceptionally keen. They could be ferociously aggressive. Prior to springing, they would bare their teeth, shift their weight forward and switch their tails rapidly from side to side. Once engaged in battle, they were vicious fighters. Pioneers learned never to underestimate a wildcat.

Henry T. Bascom, a circuit rider on the Mad River Circuit in 1814, was attacked and injured by a wildcat as he rode through the wilderness. For the remainder of his life he bore a vivid red scar on his right cheek where it had clawed him.[20]

Near Van Wert was a region where wildcats were present in unusually large numbers. Men in the community thought it grand sport to chase the creatures and watch their hounds fight with the enraged animals. Thaddeus Gilliland spoke of a particular hunt when a wildcat, time after time, evaded the exasperated hunters and their dogs. Not willing to admit defeat, they called for the assistance of a neighbor who had a tremendously powerful bulldog. When the bulldog not only caught the wildcat but also crushed its shoulder and killed it, the men were thoroughly annoyed and insisted the dog be taken home "as he was spoiling the fun. . . . Later in the day one of the hounds and a wild cat met. The cat and hound both reared up, clinched and rolled on the ground together and the hound would have soon been torn to pieces had not two other hounds come to his relief."[21]

The following story was found in a history of Miami County. "Mrs. Davis was a widow with three children and occupied a home in the wild wood region on the west side of the Miami River. About the only serious annoyance and drawback to peace was the immense number of wildcats which prowled through the woods and decimated the poultry. Mrs. Davis at last constructed, with much labor, a closed shed within which her poultry were nightly housed. This worked well for a season, but one evening a commotion in the hennery informed her that the depredators were again at work. Hastily, seizing an axe in one hand and carrying a light in the other, she hurried to the scene and two wildcats were found feasting sumptuously on her plumpest pullet. The banqueters were evidently a mother and her well grown son, whom she was instructing in the predatory art and practice. The younger animal clambered to the hole where it had made its entrance and was about to make a successful exit, when the matron, setting the light on the ground, struck the animal with the axe, breaking its back and bringing it to the ground. Without a moment's warning, the mother cat sprang upon the widow and fastening its powerful claws in her breast, tore savagely at her neck with its teeth. The poor woman, shrieking with terror, strove with all her might to loosen the animal's hold, but in vain. It tore and bit as if nothing would appease it but the luckless victim's death. Mrs. Davis would doubtless have fallen prey to its savage rage but for a happy thought which flashed across her mind in her desperate straits. Snatching her light from the ground she applied it to the hindquarters of the wildcat. The flame instantly singed off the fur and scorched its flesh. With a savage screech it released its hold and fell to the ground where she succeeded in dispatching the creature. It proved to be one of the largest of its species, measuring nearly three feet from its nose to the tip of its tail and weighed over thirty pounds."[22]

William Dean Howells, in his book, *Stories of Ohio,* told about the experiences of the white boy, James Smith, who had been captured in 1737 in Franklin County, Pennsylvania by Delaware Indians. Taken to a village on the Muskingum River in Ohio, he was adopted into the tribe and lived among them for four, generally pleasant, years. James remembered being with some of his Indian family during a period of famine when they came close to starvation. At one point, he said, they made a broth from fox and wildcat bones the buzzards had picked clean and left lying near their camp.[23]

Rev. James B. Finley left an account about the year he was twenty-one years old and he and several companions spent nearly the entire winter hunting. On one occasion they had traveled for two days without finding any game and their store of provisions was exhausted. A member of the party killed a wildcat which they hastily boiled and ate. Rev. Finley wrote, "I shall always think it was the toughest meal I ever ate."[24]

****

An 1878 newspaper carried a report about a lion in Champaign County. Residents in the northeastern part of the county had sighted a lion and the community was alarmed. Women, children and small pets were kept indoors. Doors and windows were closed and locked. Twenty-four valiant men from Champaign and Logan counties gathered for a lion hunt. Weapons in hand, they scoured the countryside for half the day. Excitement ran high. When the lion was found, it was discovered to be nothing more than a large, yellow dog that belonged to a man well known to them all.[25]

****

Panthers and wildcats, both members of the large cat family, scientific name, Felidae, once prowled over much of the North American continent. As time unfolded, their natural habitat and their prey disappeared and so did they. Ohio's rapid development in the 1830s contributed immeasurably to their decline. By 1850, the population was nearly two million. By that time the cats were gone and only memories remained of the era when "the scream of the panther and the squall of the wildcat could be heard night after night."[26] Names like Panther Creek, which is a small stream in Hardin County, and Lake Erie, "The Lake of the Cats," one of the five Great Lakes, sometimes evoked thoughts of earlier times.

Lake Erie, which is the two hundred and forty mile long lake that forms part of the international border between the United States and Canada, derived its name from a tribe of Iroquois Indians who lived on its shores in days-gone-by. Known as "The People of The Panther," or "The Cats," they called the body of water "The Lake of the Cats," or "Erieehronons." Anglicized, the word evolved as "Erie."

The Iroquois Indians and the feared but elusive panthers have been consigned to pages in Ohio history. And wildcats?

The Ohio Division of Wildlife recently reported more than two dozen verified sightings and eighteen unverified sightings of wildcats.

Chapter 3: PIONEERS: "PAINTERS," AND WILDCATS

1. Rev. David Zeisberger, *David Zeisberger's History of the Northern American Indians,* ed. Archer Hulbert & Wm. Schwarze (1910; reprint, Lewisburg, PA: Wennawoods Publishing, 1999), p 59.
2. Stillman C. Larkin, *Pioneer History of Meigs County* (Columbus, Ohio: Berlin Printing Co.,1982), p 65.
3. W. H. Beers, *History of Madison County* (Chicago: W. H. Beers & Co., 1883) p 992.
4. ——, *History of Darke County* (Chicago: W. H. Beers & Co., 1880), p 417.
5. ——, *History of Montgomery County,* (Chicago: W. H. Beers & Co., 1882), p 311.
6. Rev. David Zeisberger, *David Zeisberger's History of the Northern American Indians,* ed. Archer Hulbert & Wm. Schwarze (1910; reprint, Lewisburg, PA: Wennawoods Publishing, 1999), p 60.
7. W. H. Beers, *History of Madison Count* (Chicago: W. H. Beers & Co., 1883), p 304.
8. Henry Clay Alder, A *History of Jonathan Alder: His Captivity and Life With The Indians,* 1870, ed. Larry L. Nelson (Akron, Ohio: University of Akron Press, 1999), p 73.
9. William Steele, *The Old Wilderness Road, An American Journey* (Orlando, FL: Harcourt, 1968), pp 92-93.
10. *Urbana (Ohio) Daily News,* February 14, 1877.
11. W. H. Beers, *History of Brown County* (Chicago: W. H. Beers & Co.,1883),p 269.
12. Harlan Hatcher, *The Buckeye Country* (New York: G. P. Putnam Sons, 1947), pp 5-8.
13. R. S. Dills., *History of Fayette County* (Evansville, IN: Unigraphic, 1881), p 310.
14. W. H. Beers, *History of Miami County* (Chicago: W. H. Beers & Co.,1880), p 413.
15. *St. Paris (Ohio) New Era,* August 29, 1878.
16. Henry Howe, *Historical Collections of Ohio,* vol. 1 (Norwalk, Ohio: Laning Printing Co.,1896), p 820.
17. 1890 obituary for Margaret Ward Banes (n. p.).
18. Warren Everhart, unpublished manuscript, 1956.
19. Ray Crain, "Bits of Frontier History," unpublished manuscript, 1995.
20. Frank Siedel, *The Ohio Story* (Dayton, Ohio: Landfill Press, 1950), p 63.

21. Thaddeus Gilliland, *History of Van Wert County* (Chicago: Richland & Arnold, 1906), p 140.
22. Miami County Ohio Sesquicentennial Committee, *Miami County History* (Columbus ,Ohio: F. J. Heer Printing Co., 1953), pp 103- 104.
23. William Dean Howell, *Stories of Ohio* (Cincinnati, Ohio: American Book Co.,1897), p 54.
24. James B. Finley, *Autobiography of Rev. James B. Finley or Pioneer Life in the West* (Cincinnati, Ohio: Cranston & Curts, 1853) ,pp 147-148.
25. *St. Paris (Ohio) New Era* , April 4, 1878.
26. W. H. Beers, *History of Montgomery County* (Chicago: W. H. Beers & Co:1882), p 296.

EARLY OHIOANS AND "BARS"

Chapter 4

Bears, or "bars," as the word oftentimes was pronounced, were critters the pioneers treated with great caution, especially when the bears were hungry or their anger was aroused. Females were particularly short-tempered and ferocious when they felt their cubs were being threatened. Bears' lumbering gallop-like gait could be misleading for the large creatures were capable of moving surprisingly fast—up to thirty miles an hour. Although they had poor eyesight and were somewhat less truculent than panthers, they were, in fact, more troublesome. Bears even were known to enter people's cabins in broad daylight and help themselves to fresh meat and to milk that had been left in open containers.

While their favorite habitat was in the timberline along streams, they also wandered far beyond these confines. Male black bears usually ranged over an area of thirty to forty square miles. Female bears, on the other hand, often spent their entire life in an area no larger than ten square miles.

Like wolves, they found early Ohioans' livestock easy prey and made frequent raids on domestic hogs. Their voracious appetite for pork made keeping swine a never-ending challenge.

Seth Hawkes, an early resident in Crawford County, had a hair-raising escape from two bears that were in the process of seizing one of his pigs. He'd left his hogs in the woods to forage and when he heard one of them squealing, went to investigate. Because of a massive log lying between him and the squealing pig, it was not until he peered over the top of the log that he spied the two bears. They saw Hawkes at the same time and at once gave chase. He raced to a small tree and leapt onto a lower limb, barely in time to escape their claws as they lunged for him. Both bears tried unsuccessfully to climb the tree, their claws gouging deep cuts in the bark. Again and again they would retreat a few yards and, turning around, run to the tree and vault upward. One especially persistent bear continued its efforts to climb the tree, reaching high enough to leave teeth marks on the limb the frightened man, only seconds before, had vacated for a higher limb. Hawkes "began hallooing loudly for assistance. After about half an hour Rudolphus Morse, who had been apprised by Mrs. Hawkes of the dangerous situation of her husband, appeared upon the scene, whereupon the bears, whose fury had spent itself, apparently realizing that it was against such odds about the ownership of the hog, shambled off through the woods as fast as their feet could carry them." Mr. Hawkes shakily came down from the tree.[1]

In 1819, a thoroughly disgusted gentleman who lived in Copley township in Summit County was fined one dollar for violating the law because he killed a bear on the Sabbath. Early one Sunday morning when awakened from a sound sleep by a disturbance in his pig pen, he hastily dressed, grabbed his rifle from the rack above the door, and rushed outdoors where he saw a bear disappearing into the forest clutching a loudly squealing pig. He shot the bear and retrieved his pig. Although he argued the deed not only had saved his pig but also had rid the community of an annoying predator, none of this was taken into accord because, technically, he had been hunting on the Lord's Day. He paid the fine and whether coincidental or not, soon thereafter immigrated farther west and joined the Mormons.[2]

Henry Howe, in his *Historical Collections of Ohio,* related the experience of two women who were awakened by shrill squeals from the pigsty where a bear was attempting to get inside the enclosure. The women tried to frighten it by screaming and waving their arms. When that had no effect, they tossed firebrands and, again, it was to no avail. Although neither woman ever had fired a gun, they decided that was their only alternative. They had heard their husbands explain that it required two fingers of powder to fire a rifle. Measuring what they thought constituted two finger-lengths of gun powder, they poured it into the muzzle of the rifle. While one of the women nervously held a torch, the other pointed the gun and fired. Their misconception of the amount, "two fingers" was disastrous. The gun exploded. "The ladies both fell prostrate and insensible, and the gun flew into the bushes." The bemused bear was unharmed but made a speedy exit.[3]

A similar incident occurred in Ashtabula County. Mrs. John Austin, who was alone with her little children, heard a commotion in the pig sty and found a bear in the act of carrying off their only hog, a sow on which they were depending for a litter of pigs. "First hurrying her little children up a ladder into her chamber, for safety, in case she was overcome by the animal," she dashed outside with a rifle. The bear heard her running through the yard. He dropped the sow and, raising himself to full-height, turned to face her. Six feet away, with knees visibly shaking, and heart palpating wildly, the woman knelt and rested the gun on a rail of the fence to steady her aim. She pulled the trigger; nothing happened. The gun misfired. The bear stood inquisitively watching her while in desperation she pulled the trigger over and over again, and each time the gun clicked but misfired. After a few moments, the bear "dropped down on all fours and, leaving the hog behind, retreated to the forest and resigned the field to the woman."[4]

When Thomas Shepherd rode to Gallipolis to replenish his dwindling supply of ammunition, he left his wife, Polly, to tend the chores. During the night a bear gained access to a pen where their calf was kept. Awakened by its bawling, Polly armed herself with firebrands and boldly confronted the bear.

Waving the firebrands in front of her, she drove the beast out of the pen and up into a tree. To prevent its escaping, she built a fire under the tree and replenished it throughout the night until help arrived in the morning.[5]

Mrs. Levi Stedman of Meigs County also used firebrands to frighten a bear away from their livestock but, unlike Polly Shepherd, did not hazard going outside the cabin. Instead, she opened the door and hurled fiery pieces of wood toward the bear until it sped off into the forest."[6]

After retiring for the night, the Widow Atkinson and her daughter heard noises in the yard for which they could not account. "The mother sprang out of bed, and, adjusting her clothes, took down the rifle, which hung over the door, and after examining, cautiously, the load, and priming, she carefully opened the door and stepped into the yard. She looked in every direction but could see nothing. She walked around the house, examining every corner, but still nothing was to be seen. She also passed around the cabin of the servants but could not detect the cause of the noise. She returned into the house and set down the gun. Her daughter had arisen and dressed herself, and was stirring up the fire. She assured her mother that there must be some wild beast about the house, as she distinctly heard its footsteps while she was out. Upon this, the mother resolved to go to the cabin and wake the servant Dan, a bold and fearless negro. She accordingly went and waked him and told him to get up, which he did as speedily as possible, and after putting on his clothes, he came out armed with a heavy stick. She directed him to go round the house one way and she would take the other. Before they had proceeded ten steps, Mrs. Atkinson was seized by a huge bear. The negro immediately made a blow with his club at the head of the animal and stunned it so that he let go his hold; but soon recovering himself, he commenced his attack on Dan, who kept up a running fight till he reached his lodge, which he did, and, slamming the door, roused all the inmates.

Having thus cut off his pursuit, the bear directed his course toward Mrs. Atkinson and, just as she was entering the door of her house, caught her by her dress, and drew her toward him. At this critical moment the click of a gun-lock was heard, which was instantly followed by the sharp crack of a rifle; the bear relaxed his hold, doubled up, and rolled over at her feet, in the last struggle. An unerring aim had sent a ball through his heart.

"The daughter was a witness of the conflict in the yard; but it was too dark for her to see to shoot, without the light of her fire, and whenever the enemy came within its range, his life paid the forfeit. This intrepid act doubtless saved the mother's life."[7]

Some bears developed a proclivity for the farmers' crops. A woman in Van Wert County saw that a bear was eating corn in their corn patch and was resolved to stop it. Although crippled and no longer able to walk, it was reported that the indomitable lady dragged herself on her hands and knees for over two hundred yards in order to get close enough to shoot with some degree of accuracy. With much difficulty she positioned herself, fired, and killed it on the first try. The bear, totally absorbed in pulling off and eating the ears of corn, had been oblivious to the woman's approach.[8]

At an annual "pioneer meeting" in 1879, held in present-day Christiansburg, a Mr. Anderson was called upon to relate some of his memories of by-gone-days. He "remembered well when wolves and panthers were plenty in this country." He "recalled when a lady member of the family shot a bear which had attacked a child."[9]

Women often became adept at using firearms. A husband in Fayette County said his wife was an excellent shot. "She can bring the head off a chicken, off-hand, with ease."[10]

Ann Bailey, who rode with a rifle over her shoulder and a tomahawk and a butcher knife in her belt, was a twice-married, portly, masculine-looking woman who used profanity freely and ordinarily was slightly intoxicated. Known locally as "Mad Ann" because of her unconventional lifestyle, she seldom, if ever, wore a gown but dressed in buckskin breeches over which she donned a man's coat. Greatly admired in Gallia County for her prowess, it was said she rarely missed a shot at a bear.[11]

When James Finley was hunting in the woods one spring morning, his dogs unexpectedly startled a female bear. He shot and wounded the bear but she attacked and began mauling the dogs, killing one of them. Finley raced close to her and drove his knife into her side. She thereupon abandoned the dog and, turning to the man, caught him by the leg. "I felt that scarcely anything short of a miracle could save me. Already, I could see her wide, distended jaws ready to devour me. The dogs, though wounded, recommenced the attack, and succeeded in pulling her off, and thus saved me from death. Being released, I succeeded in killing my enemy."[12]

Thomas Moorman and his son, Chiles, with their pack of dogs, set off through the forest one frosty morning. The dogs found a scent and raced ahead where they kept a large male bear at bay until the man and the boy arrived. While it was distracted by the dogs, Chiles ran behind the bear, intending to hit it over the head with a club. He missed his aim, however, when the bear suddenly whirled around to face him. Horrified, Thomas saw it grab his son's leg and angrily begin biting it. Thomas raised his gun to shoot but boy and bear were thrashing on the ground so violently it was impossible to shoot without endangering his son. Watching as the bear savagely attacked the boy, he grew frantic. Trying to hold the gun steady, he waited for a chance to fire. At length, he took aim through Chile's legs, breathed a fervent prayer, and pulled the trigger. The bullet struck the bear in the head and instantly it fell with a thump. Although badly wounded, the boy eventually recovered.[13]

Brothers William and Jeptha Hecox treed a half-grown bear. While Jeptha sped home for a gun, William and the dogs stayed near the tree to try to hold the quarry where it sat on a branch. In spite of the frenzied barking of the dogs and William's shouting and waving a stick, the bear backed down the tree where the mettlesome dogs nipped at and tormented the young animal as it swatted them with its paws. While its attention was focused on the dogs, William picked up a heavy pine knot and, slipping around to its back, struck it a mighty blow on the head, smashing its skull.[14]

"William Cloe, then a boy about sixteen years old, called the dogs one evening and started in search of the cows. The dogs left his side and he soon heard them barking furiously at some animal that had turned at bay. He hurried forward and saw them standing guard over a large hollow log, and from their cautious movements, he knew they were confronted by an animal of which they were afraid. He stole cautiously forward from the rear, and, peering under the log, saw the huge paws of a bear. The boy was without a gun but, determining to attack the bear at all hazards, he armed himself with a heavy club and resolutely approached the log. While the attention of the bear was diverted to the dogs, which, emboldened by the approach of the boy, had renewed the attack with great fury, he seized it by the hind leg and pulled it from the log. Before the animal could recover its feet, the boy dealt it a terrible blow across the head, repeating the act again and again, until life was extinct. When the excited boy returned home without the cows and related his adventure he was not believed until the dead bear was seen."[15]

A young boy in Logan County was sent to find the cows. He had not gone far when he noticed a bear standing with its paws on a log, watching him intently. Without stopping to consider the possible consequences should he miss, he raised his rifle and shot the bear through the middle of the forehead. His incredulous family accused him of fabricating a tall tale.

"But, father, just come and see. He went and there was the bear, a very large animal, weighing nearly four hundred pounds, lying beside the log." It had been shot exactly where the boy said it was.[16]

A seven year old boy in Logan County became separated from his two older brothers while they were picking gooseberries. For eight days, unable to find his way home and subsisting only on berries, he wandered in the woods. At night, he said, he often slept in a tree; at other times he slept on the ground. On one occasion, two wolves approached him but did no harm. Later, a "large, black woolly dog came up to him and he put his hands on it and petted it." The "large, black woolly" animal had been a bear.[17]

Two sisters were herding cattle one afternoon when the younger, four-year-old, child was left to stand watch by an open gate. When the older sister returned, driving the cattle in front of her, the four-year-old was gone. Alarmed, she pondered what she should do. Then, hearing a child's faint cry and thinking it surely must be her sister, she followed the sound and located the little girl in a totally different part of the woods. Inquiring how she'd gotten there, the child explained she had been carried by a "big, black dog and that he sometimes carried her in her arms and sometimes in his mouth." From the pungent odor on her clothing and from the description of the animal, there seemed no doubt the "dog" had been a bear.[18]

An eleven year old girl who was spending the summer months with relatives often drove the cows to and from pasture. She enjoyed the stillness of the woods and watching the birds and squirrels play among the limbs in the trees high overhead; that is, she enjoyed it until the day she had a harrowing escape from a bear. Not far into the woods, the cattle moving slowly ahead of her, she heard a growl and turned to look. There was a large black bear coming rapidly toward her. She shrieked and began to run. The cattle scattered in all directions. The bear easily gained momentum on the terrified girl and, realizing there was no chance of outrunning it, she clambered up the closest tree, a remarkably small sapling. Paralyzed with fright, she clung desperately to the trunk. No sooner had she climbed as high as she could when the bear hit the tree with such force she very nearly lost her grip. The tree swayed from side to side as the bear stood on its hind legs and angrily clawed at the trunk. Men hunting nearby heard her screams and came to search for the cause. They killed the bear but not until repeatedly reassured the bear truly was dead could the still-trembling child be persuaded to slide down the tree.[19]

There was a short item in a Montgomery County history about emigrants from New England whose young children were unfamiliar with much of the wildlife in Ohio. Their first sighting of a bear cub was one of misidentification. One day shortly after arriving, the astonished children "reported to their father that they had seen a monster black cat run up the tree."[20]

In 1800, twelve year old David Murphy and a friend were paddling a canoe across the Ohio River in a canoe. They noticed a large animal, which proved to be a nearly-grown bear, swimming toward them. They watched in amazement as it swam to the canoe and, obviously exhausted, struggled to pull itself over the side. With one last heave, it fell into the boat and lay between the two boys. "The boys, of course, were very much frightened, but, nevertheless, continued paddling their canoe to the landing. The moment they touched the shore, bruin sprang out and disappeared."[21]

In 1799, George Cochran homesteaded on a fertile piece of land where the Scioto River empties into the Ohio River. "One day he spotted a black bear swimming in the Scioto. Although his gun was in his cabin some distance away, Cochran did not hesitate to push off in his canoe into the murky current in hot pursuit. His strategy was as simple as it was ill-conceived. He planned to keep prodding the bear into deep water until it drowned, when he would recover its dead body." The tenacious animal, however, possibly sensing its peril at being forced toward deep water, turned and swam to the canoe. "As the angry animal climbed into the bow of the canoe, Cochran jumped out of the stern. He then stood chest-deep in the Scioto, sadly watching his valuable craft with a bear aboard disappear downstream."[22]

Early in April of 1791, a group of people, primarily officers and their wives who were stationed at Marietta, were being rowed across the Ohio River. "They discovered a bear swimming across the Ohio River. First the men fired at it without effect. The canoe then ranged alongside when a Col. Battelle seized it by the tail and when the bear attempted to bite his hand, he raised his hind parts, throwing his head under water and thus escaped his teeth. One of the men killed it with an axe. He weighed over three hundred pounds and afforded several fine dinners to his captors."[23]

Reason Francis, in 1805, "was one day wending his way homeward on horseback through the dense forest, when he discovered a large bear, which he decided to give chase. The thought was executed by putting his horse under a good speed; but, after a long and continuous chase through the woods, and his horse being almost exhausted, the bear struck on a trail, or path, which led by the pursuer's house, standing then on the east bank of Little Darby (Madison County). Down the path the chase continued, and when passing by his house, he succeeded in getting his dog to pursue the animal, which soon resulted in treeing bruin near the creek. The dog was very vicious, and, when the bear ascended the tree, he fastened his teeth so firmly in the bear's ham that the bear carried him up the tree. Upon reaching the first limb, which chanced to be one partly decayed, the bear hoped to rest and free himself from his enemy; but alas! the limb broke, and down came dog and bear, the latter seizing the dog in his squeezers and making him terrified, when Francis, with his tomahawk, came up and buried it in the skull of his victim, releasing his faithful dog. The horse had been so completely exhausted that he soon afterward died from the effects."[24]

Captain John Minter, who lived near Radnor, Ohio, was hunting alone when he shot a bear. He reloaded the rifle, as was his custom, and presuming the bear was dead, approached and touched it with the muzzle of his gun when, unexpectedly, the bear jumped to his feet and reached for his attacker. The surprised man fired pointblank at the bear and then struck it over the head with his gun. The gun broke with the impact and flew into the air. With a downward thrust of its paw, the enraged bear knocked aside a hatchet Minter had grabbed from his belt. Frantically, Minter unsheathed his knife and in desperation began stabbing the bear. The two grappled furiously, rolling back and forth over the ground until, with a final jab of the knife, the bear died. Streaming blood from head to foot, his clothing in tatters, his back, arms and legs gashed to the bone, Minter slowly inched his way home. The morning after the battle, friends found the bear's carcass and reported there were signs that the struggle had taken place over an area of at least half an acre. It required months of recuperation but eventually the Captain did recover. However, he carried the disfiguring scars for the rest of his life.[25]

A different sort of adventure was said to have occurred in Williams County. While searching for bear cubs he knew to be somewhere in the neighborhood, John Gillet paused to rest against a hollow tree stump. It was a tall stump, approximately twelve to fourteen feet high. As he stood there trying to decide which direction to try next, he heard sounds that made him realize the cubs were inside the very tree against which he was leaning. There was no opening at the base of the tree but he noticed claw marks on the bark and concluded the opening must be at the top. Accordingly, he climbed an adjacent tree, looked down into the hole and saw two tiny cubs. In his excitement, he impetuously jumped into the den and tied the cubs' feet and mouths so they neither could scratch nor bite. He secured them to his belt and then discovered there was no way for him to get out. Inside the tree stump, "the hollow was bell-shaped, larger at the bottom than at the top and so broad I could not put my back against one side, and my feet and hands against the other, and crawl up crab-fashion." There was nothing on which to stand, no protrusion of any kind on which to climb or pull himself up.

Soon the mother bear could be heard scurrying up the tree. As she lowered herself into the den, Gillet, terror stricken now, thought of the one thing he might try. Feverously, he reached up and gripped the bear's tail with his left hand. With the right hand, he plunged the blade of his knife into her hip with all the force he could muster, at the same time yelling as loudly as he could. The bear catapulted out of the stump, Gillet hanging on for dear life. She sped down the tree and into the forest. Gillet sped in the opposite direction and toward home, the cubs still attached to his belt. The cubs he later sold for five dollars apiece.[26]

In 1801, Jacob Wise discovered a bear's den in a cliff. Hoping to capture the cubs inside, he posted a friend at the entrance to give the alarm in the event the mother bear should return. Acknowledging the strategy was foolhardy at best, Jacob then crept on hands and knees into the cave where he found two bear cubs and began tying their feet together. Like a thunderbolt, the female bear suddenly burst from the woods and swept past the startled man on guard. Roaring, and with curved claws extended, she dived into the den.

There fortunately was a second, more narrow opening at the side of the cave and Jacob lost no time sliding through it. Outside the cave meanwhile, his friend "seized her by the hinder parts" and screamed for Jacob to run for his life. Joining forces again, both men cocked their rifles and between them killed the mother bear and carried the cubs home with them.[27]

In Fayette County, Madison Loofbourrow and an Indian friend who were hunting together, were attacked by a mother bear whose cub they had wounded. The enraged bear crashed into the Indian and knocked him to the ground where she began biting him on the back, head and neck. His companion, briefly struck dumb with surprise, took aim with his rifle and dropped her with a single shot. While the two men were examining the seriousness of the Indian's wounds, a second bear cub ran from the bushes. They captured the little fellow and kept it for a pet until it grew too large to manage.[28]

Several young men in Washington County proposed to spend the winter months collecting furs to sell. They concentrated on bears, locating their winter retreats and either shooting or stabbing them to death. Early one morning they spied an easily accessible cave on the side of a hill and Samuel Brown, a member of the group, was elected to do the honors. With gun held in readiness in front of him, he entered the cave, found the bear and shot it. He'd succeeded only in wounding it however, and as it lunged to its feet, Samuel fully expected to be attacked. As he called a warning to his friends outside, he lay face down on the floor of the cave and prepared for the worst. The bear paid not the slightest attention to the intruder in its wild dash to leave the cave and, intent only in escaping, trampled over him. The hunters outside the cave had their guns trained on the opening to the cave and killed the bear as it emerged. Fearfully, they turned their attention to Samuel who still was inside. Even as they watched, Samuel came crawling from the depth of the cave, covered with blood from head to toe—blood from the wounded bear.[29]

David Murphy and his family, in 1824, were considered the first settlers in Putnam County. Once asked to write an account of his early life on the Ohio frontier, David, who was a man of few words, left the following, terse sketch: "Erected a cabin of poles; ran out of provisions; none nearer than Fort Findlay; out also of rifle balls; recollected where I had shot a ball into a tree; hunted the tree, cut out the ball; recast it, and seeing a bear on the limb of a tree, took aim at the bear—a trying moment—killed the bear."[30]

Jonathan Alder remembered a predicament that, although not amusing at the time, became so at its retelling. Having decided to spend the winter months in the Darby Plains, in present-day Union County, and hunt for whatever game might be available, he was traveling along a trail when he saw a black bear eating acorns in the top of an oak tree. Its enormous weight caused the tree to sway with the creature's every movement. Alder sat and watched, entertained by the comic performance. All at once, the bear slid down the tree and began eating acorns that had fallen to the ground. "In a few moments, without the least warning, it started in my direction as rapidly as it could pace. I had my gun resting on my lap, and saw it would run right over me! I had no time to get up and out of its way so I raised my gun and fired as soon as possible, as it was within a few yards of me. I shot it through the brain and it fell across my lap! I was badly scared for I did not know that I had killed it. I kicked and struggled to get from beneath it; but it was so heavy that it took some time before I could get out. I had shot it so dead that it never kicked and laid like a log upon me, all of which, had I not been so frightened, I might have observed. When I finally got out, I was so much exhausted from the fright and effort, I could scarcely stand. I camped at the nearest water, skinned my bear and feasted upon its flesh, which was excellent, for two or three days. So I was partially paid for my scare and trouble after all."[31]

Oak trees comparable to the one in which Jonathan Alder watched the bear feasting on acorns were commonplace in the thick Ohio forests. They grew to tremendous heights. Reverend James Smith wrote about the large, beautiful oak trees that could be found growing throughout the state. Some, he said, measured four feet in diameter at the base and stood straight as an arrow without a single branch for seventy feet. From that point to the top of the upper branch they stood an additional seventy feet. He saw men cut down these magnificent trees for the sole purpose of killing a bear; other trees, he noted, were set on fire merely to dislodge a lone bear.

A slender sapling, however, not a stately oak, better served the purpose for ten year old Thomas Ewing and his seventeen year old brother, George, who, in 1799, traveled together to the grist mill. Each led a horse on which to transport the bags of ear corn and, on the return trip, the bags of ground cornmeal. George carried a flintlock rifle and part way home decided to look for game. He handed his horse's lead strap to Thomas and began scouting ahead. Shortly, gun shots rang out. A few minutes later, George, tightly gripping his gun, rushed back down the trail closely pursued by a bear. Following the bear, barking and snapping at its feet, came the dog. Suddenly, George deftly veered to one side and darted behind a tree. The bear made a counter movement and ran in the opposite direction where it scurried up a tree and sat, resting its head on its forepaws and peering down through the leafy branches at the two boys below. A shot at its head missed but struck one of his paws instead, causing him to tumble out of the tree. Landing with a thud, he began rolling toward the creek. George shot again. The now-gravely injured animal reached the water where he seized the dog with his sound paw and held its head under water. Unable to shoot again because the flint on the gun had fallen off and landed in the grass, George, determined to save the dog, started to wade into the creek to stab the bear. Thomas yanked him back. Growing progressively weaker, the bear released the dog and stumbled to the other side of the creek where he collapsed on the sand. Meanwhile, Thomas had found the flint. George quickly loaded the gun and shot the bear a final time.

Because their horses were heavily loaded with the cornmeal and could not support an additional four hundred pounds of bear, it was necessary to leave the carcass and return for it another time. With this thought in mind, they located a suitably tall sapling and "bent it down and secured the hind legs of the bear to it—cut off the top above, and with a forked pole on each side raised the huge carcass high enough to be out of the reach of wolves, and left it till next day when they went with the necessary aid and brought it home. It was very fat—had just left its winter den in the rocks and come down to the creek bottom to feed on young nettles, their earliest spring food."[32]

As a result of their all-too-frequent forays on the settlers' pig pens and on the hogs that were running loose in the forest, some people considered black bears worthy only of extermination. Still, there were others who valued them as a ready source of meat. Bears' flesh was coarse and had a relatively strong flavor but in the autumn after they had been feeding on acorns for an extended period of time, the meat was fat and sweet. Many people liked the hams; a few even relished bears' feet roasted in hot ashes. Bear meat not used for immediate consumption usually was smoked or salted.

Indians and settlers alike used bear fat in the form of oil as an important article of trade. One black bear, depending on its size and condition, could supply as much as twelve gallons of oil.

Bear oil had a variety of practical uses, not the least of which was in preparing food. A mixture of pumpkin and meat boiled together in bear oil was considered an appetizing dish. When milk was not available for preparing mush, which was a standard meal on the frontier, bear oil was substituted.

For cosmetic purposes, Native American men and women liberally anointed their hair with bear oil to make it glisten.

A man who was constructing a log cabin wrote: "The windows of the cabin were made by sawing out about three feet of one of the logs, and putting in a few upright pieces; and in the place of glass, they took paper, and oiled it with bear's oil and pasted on the upright pieces. This would give considerable light and resist the rain tolerably well."[33]

Bear grease oiled many an axle on the pioneers' carts and wagons as they journeyed to Ohio. A family who had been away from their cabin for a number of months was overheard telling friends how they had buried their iron kettle for safe keeping while they were gone. Upon their return, they had uncovered it and coated it thoroughly with bear grease so it would be ready again to use.

Bear skins were used for bedding and for floor coverings. The hides with the thickest fur were taken from bears killed in winter and these were the most highly valued for making hats and coats. Some woodsmen learned to create an impromptu coat from a whole bear skin. They would slit the hide down the middle from the chin to the tail, turn the pelt inside-out, and use the forelegs for arms. The head they used for a cap.

Throughout most of Ohio bear skins could be exchanged for goods and had a fixed market price of three to five dollars per hide. A fifty pound bag of salt, for example, might cost one large bearskin; a Dutch oven could be purchased for one bearskin; a set of knives and forks might be bartered for one bearskin; a set of cups and saucers might require two bearskins.

Bears had little innate premonition of lures and could be captured without difficulty. Steel traps were the most popular device although bears could be enticed to climb into pyramidal-shaped log pens baited with the carcass of another animal. Crawling into an opening at the top, the clumsy animals seemed unable to find their way out again. Simon Kenton, the great Ohio frontiersman, is said to have set bear traps in this way.

Other men preferred using dead falls. "A log about a foot in diameter was fastened upon the ground at a suitable place, and wooden pins were driven into holes bored on the upper side, after which the upper ends of the pins were sharpened. Another log, fully as large, was partly suspended over the lower one, and provided on the lower side with sharpened pins, as above described. A trigger was made and baited with a portion of a dead hog, and arranged in such a manner that the bear must stand directly over the lower log and under the upper to secure the meat. To get the bait the bear must necessarily pull the trigger, which would cause the upper log to fall, thus pinning the animal like a vise between the two logs, and piercing it with the sharp pins. The trap worked like a charm, and when examined at the proper time, the bear would be found dead between the logs, pierced through and through by the pins."[34]

Some people thought it less work to capture a bear by locating a "bear tree," i.e., a tree in which a bear had made a den. It could be identified by the claw marks on the bark of the trunk. Black bears, which were the smallest of all North American bears, and the kind found in Ohio, generally were excellent climbers, ascending to surprising heights. The trick was to induce them to show themselves.

One usually successful tactic was to imitate a crying cub. When the mother bear climbed far enough out of the hole to find the source of the crying, the hunter could shoot her and she would tumble to the ground. Another method involved dropping a firebrand into the hole of the tree where the bear lay. At least two men were necessary for this operation since it required one person to climb an adjacent tree and drop the firebrand into the hole and a second person stationed on the ground to shoot the angry bear as it emerged.

Smoking bears from their dens, whether in a tree, a log, or in a cave, didn't always work, a fact to which several men in 1799 could attest. One winter morning after a heavy snowfall, fresh bear tracks were seen leading to a hillside cave. Carrying guns and axes, the men decided to follow the tracks and, if possible, smoke the bear from her lair. When the smoke failed to have the desired effect, one of the men impulsively started into the opening. He squeezed through the narrow passageway and, upon entering the cave, "could distinguish the eyes of a bear and fired at them. He then made for the entrance and in the narrow passage, a bear crashed by him and almost squeezed the life out of him." Forewarned by the noisy exit, the second man was ready and waiting and shot and killed the bear as soon as it appeared. Inside the cave lay the first bear which, although quite dead, presented a problem for it was too heavy and too large to pull through the passageway. The men shoved and yanked—to no avail. They rolled it onto its side and tried to force it out but its body became wedged, penning one of the men inside the cave. Growing uneasy at his plight, the man inside the cave struggled to pull the body back toward him again so he could slip past. Working his way around the huge body, he joined his comrade outside and, together, they braided a length of twisted hickory shoots. Securing it around the bear's shoulders, they eventually succeeded in dragging it out. For a number of days, guests at a nearby tavern ate bear meat, compliments of the two hunters.[35]

From *The Journal of Jonathan Alder* comes a story that took place while he was living among the Mingo Indians, the group that had taken him prisoner and adopted him into their tribe. He related it to his son many years after it occurred.

"One day when we were well up among the hills and bluffs, some of the Indians treed a bear up a large tree, where it had gone into a hole. The Indian tomahawk was too small to cut down the tree, so we went up the side of the hill, or mountain, and selected a small poplar that we could cut and make fall against a large limb of the bear tree. The poplar was cut and fell as described, but when it struck the large tree we heard it crack, and waited some time to see if it would break and fall; but it did not so we finally concluded it would be safe to ascend it. The next thing was to get someone to climb the small tree and drive the bear out and as I was the smallest person in the company it was decided I should go up. I did not like the idea—not that I was afraid of anything, except the bear. I consented, and was instructed to climb the tree and go a little above the bear-hole and take my tomahawk and cut a stick and punch the bear out. I knew this would be a very easy job for as soon as the stick touched him he would come out. They told me that when I saw it coming, I must climb up the tree a short distance and the bear would come out and seeing me would go down the tree and they would kill it as soon as it came into view. I stripped myself for the task, fastened my tomahawk to my belt and began to ascend the poplar. After I had gone about fifteen or twenty feet, they told me to spring up and down so as to shake the tree and see if it was safe. I shook it all I could and it appeared to be solid. I continued to go up. I got to a large limb and put my hand on it to rest. I looked down and it appeared a long ways to the ground. After I had rested a few moments, I again started; but just then, the tree broke off where I was sitting and down I went with the broken tree!

"That was about the last time I remembered, until I revived, when I found myself lying on the hillside on my back. I looked up and saw an Indian away up the tree trying to put fire into the bear-hole. One of the Indians told him to come down, adding, 'We have got one killed already with that bear and that is more than all the bears in the woods are worth; we cannot spare another. You had better come down and let him go. Just then one of the Indians came to me and seeing that I was alive, hallowed back to the others that I had come to. They all gathered around me in a short time, and I never saw people so rejoiced as they were when they found I was not dead. In a short time I began to talk and answer their inquiries as to how I felt. I told them I felt very badly and did not think I would live, but they all said I would live. They said that after the tree fell they could see nothing of me; but they cut the brush away and drew me out and supposing I was dead, had taken me up on the side of the hill and laid me down.

"While I was lying there they had cut another tree against the bear tree and an Indian had gone up to put fire in the hole but this pole was too short and they had given it up and bruin had for once come off with a whole skin after a hard day's work, of a half dozen Indians and myself.

"The fall had injured my back to such an extent that I could not stand on my feet, so they made a litter for me to lie upon and when we moved from place to place, they carried me. Sometimes two, and sometimes four Indians would perform that service for me. . . .

"I did not walk a step until the following March and then very little. . . I, however, gradually outgrew the hurt and in course of time became as stout and active as if it had never occurred."[36]

Most pioneers took pride in their hunting skills and the number of game they had bagged. David Lowry and Jonathan Daniels, in Miami County, liked to tell about the seventeen bears they shot in one season.[37] An elderly acquaintance, an experienced hunter and woodsman, gleefully reminded them, however, that it was not unusual for a single person to kill as many as eighty bears in a season!

Johnny Appleseed, who was remembered along the frontier for his travels while planting apple trees and for his belief in the doctrines of Emanuel Swedenborg, was not interested in killing bears; quite the contrary. He was noted for his kindness to animals. An exemplary story that frequently was told about Johnny said he once built a fire at the end of a hollow log, expecting to spend the night there. When he found a mother bear and her cubs already inside, he extinguished the fire and moved to a different place in the woods where he slept on top of the snow rather than disturb the bears.[38]

A story, headlined as "An Enterprising Polar Bear," was printed in a history of Ottawa County. Ottawa County, near the southwest end of Lake Erie, suffers from the cold artic winds. "The winter of 1813 was especially severe. There was not a square yard of open water that anybody knew of between the islands (in Lake Erie) and the North Pole. . . . A white polar bear of enterprising spirit started south on an exploring tour until he reached the Peninsula opposite Sandusky, when he was discovered by our kind, who treated him inhospitably, set upon him and carried off his fur coat."[39]

Captured black bears occasionally were taught to dance and, with their trainers, made the circuit of Ohio towns. Because of their naturally amusing antics—standing on their heads, somersaulting over and over again, and sitting on their haunches—they were considered "clowns." An 1877 newspaper announced, "A couple of men with a dancing bear was the attraction for small boys in town last Monday."[40] In 1881, a paper reported, "A trained cinnamon bear created a great deal of amusement on our streets Wednesday."[41] But by 1886, townspeople were beginning to voice their disapproval because the animals often were cruelly mistreated. The St. Paris, Ohio, *Era-Dispatch* wrote, "Two men and a bear made St. Paris a call yesterday. It is doubtful if they will come again, as their reception was rather cool."[42]

For many years, a tavern in South Charleston kept a black bear chained in the courtyard as an attraction for patrons. Visitors teased it unmercifully.[43]

Nick Wyant of Mechanicsburg, Ohio, used a large black bear to turn the lathe in his shop.[44]

The Eagle Coffee House, a popular tavern in Columbus in the 1830s, boasted the only public bathhouse in town. "Water for it was pumped by a big black bear chained to a treadmill. One day the bear was teased by a bystander until it broke loose. Among those who broke for a place of safety was one John M. Kerr. He leaped upon a table, and in the excitement of the occasion was unconscious for several minutes that the rear part of a dress coat he had on had been torn away by the latch of a door on which he had been leaning."[45]

By the mid-nineteenth century black bears virtually had disappeared from most parts of Ohio. Today, the normally solitary animals are protected by state law and their numbers once again appear to be on the rise, particularly in the eastern and southeastern section of the state. They remain, however, on the endangered species list.

Chapter 4: EARLY OHIOANS AND "BARS"

1. Henry Howe, *Historical Collections of Ohio*, vol. 1 (Norwalk, Ohio: Laning Printing Co., 1896), p 493.
2. ——, vol. 2 (Cincinnati, Ohio: Krehbiel & Co.,1904), p 641.
3. ——, p 278.
4. ——, Cincinnati, Ohio: Derby Bradley & Co.,1848), pp 47-48.
5. Stillman C. Larkin, *Pioneer History of Meigs County* (Columbus, Ohio: Berlin Printing Co., 1908), p 42.
6. ——, p 65.
7. James B. Finley, *Autobiography of Rev. James B. Finley or Pioneer Life in the West* (Cincinnati, Ohio: Cranston & Curts,1853), pp 144-145.
8. Thaddeus Gilliland, *History of Van Wert County* (Chicago: Richland & Arnold, 1906,), p 149.
9. *St. Paris (Ohio) New Era*, August 28, 1879.
10. R. S. Dills, *History of Fayette County* (Evansville, IN: Unigraphic,1881), p 958.
11. Henry Howe, *Historical Collections of Ohio,* vol. 1 (Norwalk, Ohio: Laning Printing Co.,1896), p 678.
12. James B. Finley, *Autobiography of Rev. James B. Finley or Pioneer Life in the West* (Cincinnati, Ohio: Cranston & Curts, 1853), p 160.
13. R. S. Dills, *History of Greene County* (Evansville, IN: Unigraphic, 1881), p 802.
14. Stillman C. Larkin, *Pioneer History of Meigs County* (Columbus, Ohio: Berlin Printing Co., 1982), p 65.
15. Henry Howe, *Historical Collections of Ohio,* vol. 1 (Norwalk, Ohio: Laning Printing Co., 1896), p 493.
16. Joshua Antrim, *History of Champaign and Logan Counties, From Their First Settlement* (Bellefontaine, Ohio: Press Printing Co.,1872), p 240.
17. *History of Logan County and Ohio* (Chicago: O. L. Baskin, 1880), pp 473-474.
18. *Mechanicsburg (Ohio) Review*, September 1, 1870.
19. *Urbana (Ohio) Citizen and Gazette*, October 2, 1899.
20. W. H. Beers, *History of Montgomery County* (Chicago: W. H. Beers & Co., 1882), p 6.
21. Nelson Evans, *History of Adams County* (West Union, Ohio: Emmons Stivers,1900), p 593.
22. Erwin Bauer, *Bears in Their World* (New York: Outdoor Life Books,1985), p 16.

23. S. P. Hildreth, *Biographical and Historical Memoirs of The Early Pioneer Settlers of Ohio With Narratives of Incidents and Occurrences in 1775* (Cincinnati: Derby & Co.,1852), pp 352-353.
24. W. H. Beers, *History of Madison County* (Chicago: W. H. Beers & Co., 1883), p 633.
25. Henry Howe, *Historical Collections of Ohio* (Cincinnati, Ohio: Derby, Bradley & Co., 1848), p 572.
26. ——, *Historical Collections of Ohio,* vol. 2 (Chicago: Derby & Co.,1904), p 848.
27. Nelson Evans, *History of Adams County* (West Union, Ohio: Emmons Stivers, 1900), p 414.
28. R. S. Dills, *History of Fayette County* (Evansville, IN: Unigraphic, 1881), p 958.
29. S. P. Hildreth, *Biographical and Historical Memoirs of The Early Pioneer Settlers of Ohio With Narratives of Incidents and Occurrences in 1775* (Cincinnati, Ohio: Derby & Co.,1852), pp 423-424.
30. Henry Howe, *Historical Collections of Ohio*, vol 2 (Cincinnati: Krehbiel & Co., 1904), p 468.
31. Johnda Davis, *The Journal of Jonathan Alder* (Columbus, Ohio: Ohio Historical Society,1988), pp 94-95.
32. *The Ohio Frontier: An Anthology of Early Writings*, ed. Emily Foster (Lexington, KY: University of Kentucky Press, 1996), pp 101-102.
33. *Ohio Historical Review* (Columbus ,Ohio: pub/ed. Ralph S. Estes, n. d.).
34. Henry Howe, *Historical Collections of Ohio,* vol .2 (Cincinnati, Ohio: Krehbiel & Co., 1904), p 848.
35. Nelson Evans, *History of Adams County* (West Union, OH: Emmons Stivers, 1900), p 480.
36. Johnda Davis, *The Journal of Jonathan Alder* (Columbus, Ohio: Ohio Historical Society, 1988), pp 56-58.
37. W. H Beers, *History of Miami County* (Chicago: W. H. Beers & Co.,1880), p 238.
38. *History of Logan County and Ohio* (Chicago: O. L Baskin, 1880), p 236.
39. Henry Howe, *Historical Collections of Ohio,* vol. 2 (Cincinnati, Ohio: Krehbiel & Co., 1904), p 365.
40. *(St. Paris, Ohio) New Era*, November 8, 1877.
41. ——, August 13, 1881.
42. *St. Paris (Ohio) Era-Dispatch*, March 19, 1886.
43. *Urbana (Ohio) Daily Citizen*, 1877.
44. Joseph Ware, *History of Mechanicsburg, Ohio* (Columbus, Ohio: F. J. Heer Printing Co., 1917), p 15.

45. Betty Garrett, *Columbus-America's Crossroads* (Tulsa, OA: Heritage Press, 1980), p 48.

DEER IN OHIO

Chapter 5

The first trappers, traders, hunters, and pioneers were impressed with the vast numbers of deer they encountered in Ohio. This is verified over and over again in the written accounts that have survived. A trapper in the mid-seventeen hundreds described the land between Lake Erie and the Ohio River as a place where there was an incredible assortment of wildlife, especially deer. James Hood of West Union, Ohio, in Adams County, wrote about droves of deer passing through their town every day. Mary Guthridge said in her memoirs there were so many deer in the area of Mingo, Ohio, where she and her parents homesteaded, they resembled large flocks of sheep grazing in a meadow. An early historian in Preble County wrote that before deer became gun-shy and wary of man, "The deer were so tame that as the men felled the trees and split rails, the deer would browse off the tops of the trees while the men would be cutting up the trunk."[1] Logan County resident, Samuel Warner, related that he had "killed as many as three [deer] in ten minutes, without leaving his tracks."[2] Sarah Moore, who came to Ohio with her family in 1821, remembered "You could go out an' kill a deer any time...."[3]

British-born emigrants, in particular, marveled not only at the multitude of deer but also that in America everyone was free to hunt for them. For centuries, venison in England was reserved for royalty and the aristocracy. Hunting for deer was the sport of sovereigns and to poach the king's deer could be punishable by death.

Daniel Constable, a native of England, came to America as a young man intending only to tour this country about which people were showing so much interest. Although he returned to England, he set sail again for America and this time, adopted it for his permanent home. In a letter to his parents, who remained in England, he wrote, "The quantity of game in Ohio almost surpasses belief; a hunter thinks it a very poor day's work if he does not shoot with his fatal rifles five or six deer and a bear."[4]

Deer thrived on the lush vegetation which abounded at the edge of the great forests. Over the span of a year, in good times, a fully-grown deer would eat a ton of succulent fruit, buds, leaves, mosses, lichens and the tender new growth on shrubs and trees. They were fond of persimmons and crab apples.

Because deer were creatures of habit, hunters who studied their daily routine found that they had specific bedding areas which they often used night after night. They fed in the early morning and again in the late afternoon. In the middle of the day deer rested, their reddish summer coats camouflaging them in the tall grasses along creeks and in thickets where it was cooler in the hot weather and where there was some protection from insects. They sometimes even could be found standing up to their heads in water. During the winter months, their coats turned a light bluish-gray color which blended with the pines and the branches on the leafless trees.

Men became adept at finding the timid animals in their quiet retreats and how best to hunt for them. Stalkers learned to stand as quietly as possible for deer had exceptionally keen hearing and could detect the slightest sound seventy-five yards downwind. There were woodsmen who chose not to wear buckskin clothing because the underbrush scraping across the surface of the leather created ever-so faint noises, loud enough to alert deer. Hunters also stood as motionless as possible for, although deer had excellent daylight vision and quickly could spot movement, it was more difficult for them to identify stationary objects.

Existence without the relatively steady supply of deer would have made life considerably more difficult than the hardships the men and women along the frontier already faced. If for no other reason, deer were important because of the meat they furnished. Venison was a staple food in many Ohio homes. Large hunting parties, and surveying crews relied on deer to help keep them in fresh meat. Even the animals' tongues and muzzles were considered superb eating and to keep the peace, the men tried to divide them evenly at mealtime. Meat not used immediately often was preserved in salt or "jerked," i.e., smoked or sun-dried. Cut into strips, jerky was light-weight and easily carried in leather pouches. It kept well and could be eaten at any time. It frequently was boiled with other food for a quick meal.

Venison was important enough that our pioneer ancestors went so far as to devise a method of reconstituting the meat which had spoiled. In a recipe entitled "To Recover Venison When It Stinks," the directions read: "Take as much cold water in a tub as will cover it a handful over, and put in good store of salt, and let lie three or four hours. Then take your venison out, and let it lie in as much hot water and salt, and let it lie as long as before. Then take it out and dry it very well, and season it with pepper and salt pretty high . . . ."[5]

Surprisingly little meat ever spoiled, however, for families were large and food, food of any kind, was consumed quickly. When game was scarce and times were hard, missing the opportunity to bring home a deer could be a serious matter. The name of the boy in the following incident is unknown, but it aptly exemplifies the situation. "A good-sized boy had gone fishing in the river back of Staunton (in Miami County). A deer came down to the river to drink and having his gun with him, he shot the deer. An Indian appeared out of the bushes and took the deer away from him and when he told his father about it that evening, his father whipped him for letting the Indian take the deer."[6]

General "Mad" Anthony Wayne, who succeeded Arthur St. Claire as commander-in-chief of the Northwest Army, commissioned Josiah Hunt during the winter months of 1794 to supply the officers' tables with game. Normally, he hunted for deer. Always on guard for Indians who may have heard him fire his gun, he made it a practice to approach the deer cautiously after he had shot it. He would stand with his back against a tree as he skinned the kill. Rifle leaning close to his right hand, he would skin for a short time, straighten up and peer in all directions, then resume working. Wrapping all he could carry in the slain deer's hide, he would sling it over his back and make his way back to the fort. On many occasions, it required several trips to carry all the meat and the wily hunter used a different route each trip to conceal his presence from the Indians.[7]

Indians, as a rule, liked their venison less-well cooked than the pioneers did. For instance, Half John, a Wyandot Indian who befriended a Van Wert family, preferred his deer meat cooked so that when he was eating it the blood would run out of each side of his mouth.

An Indian who stopped at a settler's home in Summit County made it understood that he wished to spend the night with them. From a deer carcass draped over the back of his horse he cut a hunk of meat and asked the woman to cook part of it for him. She sprinkled it with salt and pepper and prepared it as she customarily did for her family but when it was done, the Indian ate only a few bites and then pushed it aside. Placing his rifle, his scalping-knife and his tomahawk in a corner of the cabin, he stretched-out on the hearth in front of the fire as if to sleep. The owners of the cabin also retired for the night but were uneasy about the uninvited guest sleeping across the room from them and merely pretended to sleep. When the Indian thought that his hosts were asleep, he rose and went to the corner where he picked up his scalping-knife. The settler, still covertly watching the Indian, was seriously contemplating reaching for the gun he'd leaned against the bed when, rather than approaching the bed where the man and his wife lay, the Indian turned to the table and cut another piece of venison which he dropped onto the hot coals in the fireplace. Warming it only slightly, he picked it up with his knife and ate it. He finished eating, stretched-out again and went to sleep.[8]

Mrs. Nathan Hill of Miami County was alarmed when an old Indian, Amokee, came to the house one day. She visibly relaxed, however, when she saw he was carrying his gun over his shoulder with the barrel pointing toward the ground, a sign that he meant no harm. He walked into the cabin and asked for food. Apologizing because there was no meat, she prepared a large serving of corn pone for him. Later in the day Amokee returned, this time bearing the hind quarter of a deer which he indicated was for the Hills. Amokee "remained with them until Spring and then departed."[9]

Deer skins were as important as deer meat and trade in their hides was brisk. In most areas the hides were acceptable legal-tender. Daniel Clark, a minister in Greene County, said there were years when the only salary he received consisted primarily of buckskins.

To say that an item was worth "a buck" referred to the use of buckskins as currency. Traders often set the price for a buckskin (the skin from a male deer) at $1.00; a doeskin (the skin from a female deer) at $1.50 to $2.00. A yard of fabric might cost a doeskin. A ready-made man's shirt was worth two buckskins. A chintz shawl cost four buckskins and two coonskins. Four pounds of sugar required two doeskins plus a coonskin. A pound of tea cost three buckskins. Fifty pounds of salt cost four buckskins and two coonskins.

Hunters put in long, hard hours skinning and scraping but the profits were worthwhile when it was taken into account that a horse easily could transport one hundred deer skins at a time. The pelts one horse alone could haul would buy a great many items. A man in a single season might kill two hundred deer.

Skinning a deer was a learned art. The simplest method was to hang the slain animal head down from a limb of a tree and with a quick, steady stroke of a razor-sharp knife, slit the skin up the stomach from the tail to the neck, circling around the top of each leg and around the neck. Clutching the skin in one hand, it then could be pulled and cut loose from the flesh.

Many pioneer families tanned their own leather. It was coarse but quite serviceable. Eventually, most communities had at least one tannery whose business it was to convert raw hide into tanned leather. Professionally tanned leather was softer and finer and far more desirable than home-tanned skins.

While clothing along the frontier followed no set style, trousers, leggings, and vests of "buckskin" were worn by many of the men. When wet, the leather was notoriously cold and uncomfortable. A settler in Greene County who was immensely pleased with his new buckskin shirt and trousers, changed his mind the first time he was caught in a downpour. He fervently wished that he'd never, ever, heard of buckskin clothes. The only way he could make them wearable again was to knead them in the same manner his wife kneaded bread and then to pound them with a heavy stick.[10]

An Erie County resident reached the same conclusion about his buckskin clothing. He said that "when dried, after being wet, they were hard and inflexible; when thrown upon the floor, they bounded and rattled like tin kettles. A man on a cold winter's morning, drawing on a pair, was in about as comfortable a position as if thrusting his limbs into a couple of frosty stovepipes."[11]

An entertaining story about a young man and his buckskin pants was told in a Madison County, Ohio, history book. "A young man went to see a young lady, both of good families. The young lady had several brothers, and they kept a number of hounds to hunt coon, which was a good business in early days. Well, after the young couple had talked until time to go to bed, the young lady told the young gentleman where to retire, and, by some mistake, one of the hounds obtained an entrance into the room. The young gentleman laid his pants down on the floor; but they being made of buckskin—which were very fashionable in those early days; the hound, being hungry, ate one leg off his pants. In the morning, when he arose, he wished himself at home; but she furnished him a pair of pants to wear and sent him on his way rejoicing."[12]

Lice on deer hide clothing could be an annoyance. Indian women sometimes were observed chewing the seams of their husbands' and sons' deerskin shirts in an effort to kill the irritating parasites.

Moccasins were made from deer hide. The wearer's feet had to be tough for the soles of the moccasins were thin. One pioneer made the comment that moccasins were little more than "a decent way of going barefoot."[13] Like buckskin clothing, unless well greased with animal fat as a form of waterproofing, they absorbed water. When sleeping around a campfire, woodsmen made an effort to put their feet close to the fire in order to dry them. Moccasins were hung to dry on sticks driven into the ground near the fire.

If feet remained moist or wet too long, the flesh would begin to rot and peel, a condition known as "scalded feet." A salve prepared from slippery elm bark could be applied but the healing process was a slow one. In spite of the drawbacks, moccasins became the standard footwear in Ohio.

To help keep their feet warm and comfortable, Native Americans used deer hair to line their moccasins. Woodsmen quickly adopted the idea and found it often replaced the need for stockings.

Deer hair was said to be responsible for the name of the Scioto River. The Indians observed that deer were so numerous during the autumn when they came to drink at the river, it would be thick with their hair. As a result, they called it "Ough-scan-oto," meaning "deer" and "hairy river." The white man translated it as "Scioto."

In preparation for spring planting, it became a common practice to coat seeds with deer brains as a way of enhancing germination. Animal skins well rubbed with a sudsy mixture of deer brains and urine made the hides soft and supple."

The blood of freshly killed deer frequently was listed in the annals of frontier medicine as a cure for a variety of ills. More than one early pioneer drank deer's blood and attributed it to making him well again.

For the rattling sound they created, strings of deer's hooves tied around their ankles and knees were worn by men of the Delaware Indian nation while performing certain kinds of dances.

Although tanned deer hides were the most highly prized, settlers did utilize rawhides as well, cutting them into long, narrow, strips which, twisted or braided together, made harnesses, bridles, and ropes. By means of sturdy rawhide thongs, Indians and early hunters often carried their rifles slung over their back and when traveling long distances would use deer hide thongs to tie their guns, each one wrapped in a piece of leather for protection, to a bundle behind their saddles. Even after deerskin pants were discarded in favor of cloth trousers, large buckskin patches frequently were worn over the knees and seat. Leather leggings, nothing more than a piece of rawhide buttoned around the lower part of the legs and secured with a string below the knees, were worn as a shield against dust, mud, and venomous snakes.

Every now and then, un-tanned deerskins stretched over willow hoops with holes burned in the hides by means of a red hot poker served as primitive sieves.

Indians made the cylindrical skin from around the creatures' necks into vessels to hold liquids. Rendered bears' fat customarily was stored in this manner.

It was not unusual to use the skin of a slaughtered deer to transport honey from a felled "bee tree." The legs of the dressed skin were tied together and a pole inserted between them. Two hefty men would hoist the pole over their shoulders and carry the honey home.

The most common method of hunting for deer was with a gun but not everybody was skilled in the use of firearms. Thomas Cowgill wrote that his father, "on his way from New Jersey, when at Lancaster, Pennsylvania, purchased a new rifle, a kind of firearm he never used, but during his first winter in his new home in Logan County, when there would come a fall of snow he would take his rifle and practice hunting, and succeeded in killing a turkey or a rabbit now and then, but from that nervousness and anxiety called buck-fever, could not for a long time succeed in killing a deer. But one morning after a fall of light snow he tied a white handkerchief over his head and dressed in light-colored clothing, assimilating as near as possible to the color of the snow, put out, gun in hand into the forest immediately back of his cabin, and was not gone more than ten minutes until the family were saluted with the shrill crack of his rifle, and looking in the direction of the report he was seen running at full speed toward the cabin, with his gun held horizontally in both hands, in a perfect fever of excitement, and out of breath, and entirely speechless, thrusting the cock of his gun aimed in the faces of his family, to let them know he had killed a deer; he had to be even reminded that he must stick it, which he had forgotten under the frenzy of his buck-fever; he went immediately back and stuck it which had dropped dead from his shot, after which he was more deliberate and cool, and became a tolerably good hunter both in the chase and at deer licks, which abounded at that time in that part of the state."[14]

Deer hunters made note of the location of deer licks, or salt licks as they also were known. These were natural deposits of salt in the earth where deer sometimes congregated to lick the briny compound. If a person were patient enough, deer invariably came. It was an easy way to hunt for game.

A deer lick in Logan County was the scene of a grim battle between Joseph Curl and a fine, young, buck deer he thought he'd killed. The moment it was shot the buck had dropped to the ground and lay motionless. Pleased with his good fortune, Curl approached. Suddenly, the deer sprang to its feet and charged. Curl dodged behind a tree where the two engaged in an involuntary game of tag around the trunk until "by a lucky stab, he disabled the deer and finally killed him."[15]

A similar incident occurred in Paulding County. A hunter laid his rifle on the ground after presuming the deer he had shot was dead. He lifted the animal's head by its antlers to expose the throat in order to cut and bleed it. As he gripped the antlers with one hand, he unsheathed a knife with his free hand and was starting to bend over when, all at once, the deer began staggering to its feet. Retaining his hold on the antlers and forcing them downward, he backed toward a small tree where he pulled the deer's neck against the trunk. Around and around the tree the pair moved, the trunk always between them. At no time did the man relinquish his grip on the antlers. Both hunter and beast were exhausted when the buck momentarily ceased struggling and the man slit its throat.[16]

Deer had difficulty running in deep snow because their sharp hooves sank through the soft surface, considerably slowing their progress. Snowdrifts were particularly treacherous. Crust on top of the snow also caused them additional problems. Although their legs were strong, once they broke through the crust they would flounder. Hunters and wild animals reaped the benefit of the situation.

An account in a Preble County history states that the winter of 1817 was a particularly harsh one. "A fall of snow covered the ground for a depth of two feet, and a crust of snow formed thick enough to bear the weight of a dog but not that of a deer. The dogs of the settlers set out on a hunt for themselves, and the old settlers declare that such was the havoc made by this wholesale slaughter that 'scarcely an antler' was seen in the country afterwards. The dogs could easily pull down a deer, which, at every step, would plunge up to its body in the treacherous snow, while they could glide along at full speed over the frozen ground, barely stopping to make sure of the death of one before starting off after a fresh victim."[17]

Wintertime or summertime, deer occasionally were captured in pits that were dug along the paths they traveled to feeding grounds and to streams and rivers where they drank. Digging a pit required an enormous amount of work, however, even when a gully or a ravine was utilized. There were much easier ways of killing deer.

Hunters were known to place the head of a slaughtered deer on a stick and creep through the tall prairie grass until a live deer was decoyed within range of a rifle.

From July until September, deer habitually went to the river after dark to avoid flies and to eat the tender water plants that grew at the water's edge. Men floating silently downstream, hidden from view behind the bright glare of a torch positioned in the bow of a canoe, often hunted deer at night. The shimmering light reflected on the water and attracted the attention of the inquisitive creatures. They would raise their heads to gaze at the anomaly and, while the animals were distracted by the glare, the men could drift within shooting distance, sometimes to within ten or fifteen feet of them.

James Smith and John Brickell found deer in this manner and said it was not unusual to kill twelve deer in one night. Two brothers near Lima boasted of having killed eighty deer in a single season, oftentimes shooting six a night. On occasion, other species became the unwitting victims. As early as 1784, a few communities in the East had banned this type of nocturnal hunting for the accidental slaughter of cows and horses became too great.

Nighttime hunting also could prove dangerous for human beings. One summer evening, John and Samuel Morningstar spotted a large buck standing close to the creek and, when within range, John fired and wounded the animal. Shrieking loudly as injured deer are prone to do, the frenzied buck ran headlong into the canoe, capsizing it and dumping the boys in the water.[18]

Jonathan Alder now and again hunted in this way and said the lighted torches blinded and confused deer and they would have a tendency to walk toward the light. "I have frequently had them jump right into the canoe onto me and sink me in the creek. It is always warm weather when we fire hunt and I would very often have my leggings off and my thighs and legs all bare. The first thing I would know, I would feel a snake running across my thighs. They [like the deer] would see the light and make for it and come to the canoe and swim right up into it, and across, and out on the other side; many a one I have killed with my pushing pole when I would see him coming."[19]

Jonathan described another method he sometimes employed to kill deer. Referred to as a "fire ring," he cut a handful of long grass to make a torch and, setting it afire, would run with it "circling around and taking in three or four thousand acres." Jumping inside the ring, he often killed as many as seven deer. "When I would kill one, I dragged it into a thicket where there was no grass and scraped the leaves away so as not to have the hide injured by the fire."

The last time Jonathan made use of the tactic, "the fire began to close in on me and it burned very rapid. I could see no good place of escape. I looked me out a good piece of ground where there was no brush and when the fire began to get pretty warm, I put my powder horn under my arm and fired off my gun, then leaped. I had wrapped my blanket tight around me—head and face all covered. I could not see a particle. I was perfectly blindfolded. I turned my face in the direction that I wanted to run before I covered it. The fire was then a perfect blaze ten or fifteen feet high and I started and ran through it. The main blaze was not more than thirty or forty feet wide, but I ran about two hundred yards before I recovered. I was out of the main fire, but it was still burning and I had to run further to get entirely out of the fire on account of my powder horn. My moccasins were entirely ruined and my leggings and blanket were nearly spoiled. I hunted up the deer and skinned them. Some of them had their hair pretty well singed off, but the hides were not injured. But that was my last ring fire."[20]

According to Crawford County chronicles, Indians oftentimes killed does by imitating a fawn bleating in distress. When the doe ran to her offspring, she would be shot. Settlers in the community became concerned by the numbers of does being destroyed in this fashion and finally asked the Indians to stop. When a member of the tribe disregarded the appeal and killed yet another doe by mimicking a fawn, a group of the men confronted the guilty party. "With significant taps upon their rifles," the Indian was informed that if the act were repeated, he "would be shot."[21]

As a result of the many does that were slaughtered, there were countless orphaned fawns. Scores of the motherless animals became much-loved pets. Nearly every cabin sooner or later had its pet deer. After only a few days with a settler's family, a fawn would follow closely on the heels of its surrogate mothers. John May, a Revolutionary War veteran who landed in Marietta in 1789, raised four orphaned fawns. They were "as tame as cats," he wrote, "climbing all over [me] and sucking on [my] ears."[22]

A trapper who was passing an isolated cabin deep in the forest watched as a small boy ran out the door followed by two huge dogs and a tiny pet fawn. "The dogs seemed to be very fond of this innocent little thing. It skipped along and played round the footsteps of the child, very affectionate," he said.[23]

In order to prevent these vulnerable pets from becoming lost in the woods and from being shot by mistake as wild deer, bells generally were hung around their necks. This did not always guarantee their safety, however. A pet deer with a bell prominently hanging around its neck and feeding close to its owners' house, was shot and killed by a neighbor. When the same neighbor killed more pets in the vicinity, supposedly by accident, indignant residents cornered the man and gave him fair warning there must be no more such "accidents" or he would suffer the unpleasant consequences.

Alexander McKee, a licensed Indian trader who fought in Lord Dunmore's War alongside Simon Kenton and Simon Girty, established a trading post in McKee's Town just south of present-day Bellefontaine, Ohio. Here he built a comfortable hewn log house and in 1794 enjoyed the company of a pet deer that was allowed the freedom of the house. One morning while McKee was in the process of dressing, the nearly-grown young buck charged into the room and, with head lowered, playfully struck the man. According to one of several different versions regarding the mishap, a spike on the yearling's partially developed antlers stabbed the old trader's thigh, possibly piercing an artery because in a relatively short time Alexander McKee bled to death.[24]

While fighting, adult male deer used not only their antlers, which, when fully developed could be fearsome weapons, but also their sharp front hooves. During the rutting season each autumn, bucks needed no provocation and would initiate an attack without cause. Female deer fought to protect their fawns and for their own lives and although they had no antlers, their front hooves were a formidable means of defense.

Captain John, a well-known Delaware Indian who had befriended Madison County settlers, frequently visited and hunted with them. During his last visit to the area, he was discovered in the woods lying beside a buck deer. Both were dead. From all indications, it appeared that Captain John fatally had been stabbed by the deer's antlers. The deer, in turn, had died from injuries Captain John had inflicted.[25]

William Steele, in his book, *The Old Wilderness Road*, tells of a buck deer's charge and the furious battle that ensued. The attack was so unexpected that the hunter had no time to cock and swing his gun to his shoulder. As he leapt aside, he felt the antlers scrape across his back. Before the enraged animal could wheel and charge again, the man had his gun ready and fired. Although the bullet struck the deer, the animal merely lowered his head and rushed full tilt at him. Bracing himself, the man hit it across the nose with the stock of his gun. Stunned for a moment, the deer stood shaking its head from side to side and while so engaged, the hunter hit it again with his gun. Dropping the gun, which now had disintegrated, he reached for his knife and grabbed the beast around the neck and slit its throat. Incredulously, the buck continued to lunge and twist, trying to dislodge its tormentor. Finally, "the buck fell to its knees." The exhausted man watched "the deer slowly slump over to its side. A long deep sigh and it was dead."[26]

In 1821, two Crawford County residents agreed to a wager that each could kill more deer than the other in a day's time. One of them had an ancient flintlock rifle so battered and worn that it was held together with leather thongs. The man was, nevertheless, an experienced hunter with the old gun and by evening had killed seven deer. He also had shot and mortally wounded an eighth deer but darkness prevented his following and retrieving the carcass. Because he could not prove he had killed it on the day of the contest, which would have resulted in a tie, he lost the wager.

Deer skins were valuable items for trade. Traditionally, Indians used practically every part of the deer they killed but when they learned the hides could be exchanged for the goods they desired such as knives, cooking utensils, beads, and alcohol, they began slaughtering them by the hundreds. They would skin the animals on the spot and leave the meat to rot or for the animals of the forest to consume.

One of their discarded deer heads was credited with helping to keep two women captives from starving to death. Ted Morgan, in his *Wilderness at Dawn*, tells the true story of Mary Ingles, a white woman who was captured in 1755 and carried to a Shawnee village where she was separated from her two-year-old and four-year-old sons. Forced to work as a slave, she resolved to escape. Her chance appeared when she and another white woman were sent, unattended, to gather walnuts and grapes. Knowing their lives would be in jeopardy if recaptured, they nonetheless slipped into the woods, and headed east, toward home.

Their clothes in shreds, they tore strips of the tattered fabric to wrap their feet when their moccasins wore out. There was virtually nothing to eat during their harrowing forty-day flight through the untamed wilderness. They were eternally grateful one day to find the putrefying head of a deer an Indian hunting party had killed. Ravenous, they devoured every scrap of flesh that adhered to the skull.

Late in November and barely alive, Mary, who by now was not far from home, was discovered by former neighbors. She was nursed back to health and before long reunited with her husband. Her companion, originally from Pennsylvania, had farther to travel, but eventually found the way to her home.[27]

Nine year old Levi Perry and his brother, eleven year old Pepper, accompanied their father to a tract of land they had purchased near Delhi, in Delaware County, to build a cabin and to prepare land for crops. In September of 1803, Mr. Perry left the boys to pile the underbrush and the tree tops he had felled and returned to their former home in Philadelphia to bring Mrs. Perry and the rest of the family to their new home in Ohio. Not only did Mr. Perry find his wife seriously ill but he, also, became ill. Expecting to be back with the boys by mid-to-late October, it was, instead, nine months before the Perrys were well enough to travel.

Meanwhile, the Ohio winter was one of the cruelest on record. The cabin, which had neither a fireplace nor a chimney, nor caulking between the logs in the wall, served as little more than a rude shelter for the two little fellows. Long before the parents came, the food supply was exhausted. The boys survived on a bit of corn meal they occasionally were able to obtain from a family fifteen miles away and they caught and ate rabbits they sometimes managed to drive into hollow logs. Once they found the remains of a deer that wolves had killed. Amazingly, Levi and Pepper lived through the winter and had cleared a large part of the ground by the time their family arrived in June of 1804.[28]

Sometimes the misinformation and/or lack of knowledge on the part of new settlers was the source of much amusement. A recently arrived Irish immigrant and his family lodged briefly in Piqua, Ohio. Their sixteen year old son had not returned from the forest at suppertime one evening and the distraught mother insisted the men search for him because, she sobbed, "the opossums will kill him, and the deer will eat him." A bonfire was lit to serve as a beacon and guns were fired to attract the boy's attention. "In about half an hour" the boy appeared.[29]

The vast herds of deer in the great Ohio forests were decimated in a relatively short time after the white man discovered the region. A biographical sketch of Allan Loudenback, who resided in Mad River township, Champaign County, included the statement that in 1837 he had shot and killed the last wild deer in the county. Sighting one of the beautiful creatures subsequently became a novelty. A journalist for a 1931 Urbana Daily Citizen wrote, "A deer has been seen at various times recently in this and adjoining counties. Residents in the vicinity of Locust Grove in Clark County have reported seeing a buck deer in that locality the last few days. A few days ago some DeGraff hunters reported seeing a deer in a woods near West Liberty."[30]

Nearly one hundred years after Allan Loudenback shot the last deer in Champaign County, the white-tailed deer were reintroduced into Ohio. They made a rapid and steady comeback. By the year 2000, deer in Champaign County were estimated at the thousands. Since their reentry in the 1930s there now are more wild deer in the state than at any time during the pioneer era. Our forefathers would be amazed.

****

Although never seen in abundant numbers, elk were found in Ohio until circa 1795. They were larger than deer and had branching antlers, sometimes growing to an enormous size. Reverend James B. Finley was quoted as saying he had seen antlers so gigantic, a six foot man could pass under them without stooping.[31]

Information regarding these regal animals; however, seldom is read in early histories. An entry in Christopher Gist's first journal, dated 1750-1751, noted that Delaware Indians reported elk along the river they called "Mooshingung" ("Elk's Eye River"). "Elks were so plenty on that river and so tame the Indians could come as near as to see into their eyes so they called the river Mooskingung or "Elk's Eye."[32] White men translated the Indian word as "Muskingum," the name the river retains yet today.

David Zeisberger, a missionary for many years among the Delaware Indians in Ohio, briefly mentions elk in a book he wrote about North American Indians. He said their skin was thick and heavy and of no particular value; that Indians were not tempted to hunt for them and only shot them when necessary. He did say, nonetheless, that some Indians preferred elk skin moccasins for they felt the hide was stronger and more durable than deer skin.

    S.C. Larkin, in his *Pioneer History of Meigs County*, related a story about two boys and an elk. Brothers, George and John Warth, while searching for game in late autumn, scouted in different directions but remained within calling distance of each other. Soon after parting company, John heard a gun shot and almost immediately afterwards, cries for help. He raced toward the sound of his brother's cries and saw that a full-grown bull elk had George pinned to the ground with its antlers. Breathless from running and gasping for air, John knew he could not hold his rifle steady enough to shoot with any assurance he would hit the beast. The elk kept twisting its head, attempting to disengage his antler and try again to gore the helpless boy. John was desperate, so walking directly to the elk, he thrust the muzzle of his gun against its ribs and pulled the trigger. The bull crashed to the earth. George was bruised and badly frightened but otherwise not really hurt. One of the prongs on the elk's long antlers had pierced a fold of George's shirt and it was this which had been holding him fast.[33]

    It was said that a tribe of Indians in the West, who were called the "Copper-Heads," occasionally broke elk to ride and to serve as beasts of burden. An observer, Matthew Brayton, a one-time-captive of the tribe, explained how it was done. One of the men in the village would lasso an elk's antlers and tie the end of the rope to a tree. A second lasso secured the antlers to a tree on the opposite side and both ropes were drawn tight. Additional lines were used to anchor the hind legs, effectively restraining the animal while the man who was chosen to tame the elk would leap onto its back and clasp it around its anthers. Clamping his legs against the startled creature's heaving sides, he crossed his feet beneath its stomach to keep from sliding off. Simultaneously, his companions cut the ropes and away the elk would fly.

Following a long and exhausting run, the animal stood, utterly spent. The rider, still sitting on its back, would hold a handful of sugar or salt under its nose and slowly move the treat down to its mouth where "the taste of these articles is generally sufficient to subdue the strong will [of the elk], and to complete the work the rider puffs tobacco smoke up its nostrils. It is now thoroughly broken in, and will sit easily under a rider or follow its owner like a dog."[34]

Elk, or wabete, as they were called by the Shawnee Indians, have been reintroduced in small numbers to several areas in the State of Ohio. Unlike the flourishing deer, according to the Ohio Division of Wildlife, the program cannot yet be termed a success.

Chapter 5: DEER IN OHIO

1. *History of Preble County* (Cleveland, Ohio: H. Z. Williams & Bros., 1881), p 175.
2. *History and Biography of Logan County* (Chicago: O. L. Baskin & Co.,1880), p 473.
3. *(Urbana, Ohio) Champaign Democrat*, June 6, 1911.
4. Brian Jenkins, *Citizen Daniel (1775-1838) and The Call of America* (Hartford, CT: Aardvark Editorial Services ,2000), p 74.
5. Julianna Belote, *Compleat American Housewife 1776* (Conrad, CA: Nitty Gritty Productions, 1974), p 83.
6. Miami County Ohio Sesquicentennial Committee, *Miami County History* (Columbus, Ohio: F. J. Heer Printing Co., 1953), pp 17-18.
7. William A. Galloway, *Old Chillicothe* (Xenia, Ohio: Buckeye Press, 1934), p 218-219.
8. Henry Howe, *Historical Collections of Ohio*, vol. 2 (Cincinnati, Ohio: Krehbiel & Co., 1904), p 632.
9. Miami County Ohio Sesquicentennial Committee, *Miami County History* (Columbus, Ohio: F. J. Heer Printing Co., 1953), p 292.
10. R. S. Dills, *History of Greene County* (Evansville, IN: Unigraphic, 1881), p 424.
11. Henry Howe, *Historical Collections of Ohio*, vol. 1 (Norwalk, Ohio: Laning Printing Co., 1896), p 566.
12. W. H. Beers, *History of Madison County* (Chicago: W. H. Beers & Co., 1883), p 664.
13. R. S. Dills, *History of Fayette County* (Evansville, IN: Unigraphic, 1881), p 26.
14. Joshua Antrim, *History of Champaign and Logan Counties, From Their First Settlement* (Bellefontaine, Ohio: Press Printing Co., 1872), p 20.
15. *History and Biography of Logan County* (Chicago: O. L. Baskin, 1880),p 473.
16. Thaddeus Gilliland, *History of Van Wert County* (Chicago: Richland & Arnold, 1906), p 137.
17. *History of Preble County* (Cleveland, Ohio: H. Z Williams & Bros., 1881), p 170.
18. R. S. Dills, *History of Greene County* (Evansville, IN: Unigraphic, 1881), p 588.
19. Henry Clay Alder, *A History of Jonathan Alder: His Captivity and Life With The Indians,* 1870, ed Larry L. Nelson (Akron, Ohio: University of Akron Press, 1999), p 84.
20. ──, p 112.

21. Henry Howe, *Historical Collections of Ohio,* vol. 1 (Norwalk, Ohio: Laning Printing Co.,1896),p 492.
22. Ted Morgan, *Wilderness at Dawn* (New York, N.Y.: Simon & Schuster, 1993), p 429.
23. R. E. Banta, *The Ohio* (New York, N.Y.: Rinehart & Co., 1949), p 324.
24. Edna Kenton, *Simon Kenton-His Life and Period 1755-1836* (New York, N.Y.: Random House from the Doubleday edition, 1930), p 57.
25. W. H. Beers, *History of Madison County* (Chicago: W. H. Beers & Co., 1883), p 670.
26. William Steele, *The Old Wilderness Road An American Journey* (Orlando, FL: Harcourt, 1968) pp 85-86.
27. Ted Morgan, *Wilderness at Dawn* (New York, N.Y.: Simon & Schuster, 1993), pp 394-396.
28. Henry Howe, *Historical Collections of Ohio,* vol. 1 (Norwalk, Ohio: Laning Printing Co., 1896), p 551.
29. ——, vol 2 (Cincinnati, Ohio: Krehbiel & Co., 1904), p 723.
30. *Urbana (Ohio) Daily Citizen,* November 10, 1931.
31. James B. Finley, *Autobiography of Rev. James B. Finley or Pioneer Life in the West* (Cincinnati, Ohio: Cranston & Curts, 1853), p 79.
32. William Darlington, *Christopher Gist's Journals With Historical, Geographical and Ethnological Notes and Biographies of His Contemporaries* (1780; reprint, Pittsburgh PA: J. R. Weldin & Co., 1893), p 103.
33. Stillman C. Larkin, *Pioneer History of Meigs County* (Columbus, Ohio: Berlin Printing Co.,1982), pp 46-47.
34. J. H. A Bone, *Adventures of Matthew Brayton The Indian Captive* (Cleveland, Ohio: Harold Office Printers, 1860).

BUFFALO

Chapter 6

Although zoologists never considered North American buffalo true buffalo but classified them "bison," the name "buffalo" persisted. Ideas differ but the misnomer possibly originated in 1544 when Spanish explorer, Hernando de Soto, who actually may never have seen the animals, used the term in his journals to portray the large, hairy beasts. The author of a book entitled *The Natural History of Carolina Florida, and the Bahama Islands,* perpetuated De Soto's nomenclature and after its publication in the mid-1700s, "buffalo" came into general usage.

An early explorer who studied the animals at close range wrote that a buffalo has "crooked shoulders with a bunch on its back like a camel, . . . its neck covered with hair like a lion, . . . its head armed like that of a bull."[1] Another observer commented on the tassel-like tails. "Their tail is very short and terminates in a great tuft. When they run they carry it in the air like scorpions."[2] Someone also noted, "There is a mixture of the awful and the comic in the look of these huge animals as they bear their great bulk forward."[3]

The most disparaging account of buffalo read: "Nothing can be more revolting, more terrific, than a front view of an old bull buffalo . . . a dirty drunkard beard . . . altogether the appearance and expression of some four-legged devil."[4] Indeed, the coarse hair on a bull's throat and chin that culminates in a twelve inch long beard does contribute to the alarming illusion of a devil. Some of the 17th and 18th century artists went so far as to draw buffalos as vicious brutes with large, red, glaring eyes.

Ohio records show that buffalo, occasionally referred to as "wild beef," once roamed from the south shore of Lake Erie to far below the Ohio River. Herds of five hundred or more were not uncommon. However, relatively few of the first pioneers saw herds of that size. By the time they had arrived, herds of forty or fifty were more prevalent and by 1800, they had vanished almost entirely. Two buffalos were killed in 1800 in Jackson County. In 1803, a lone buffalo bull was killed in Lawrence County. In 1806, a settler killed a buffalo at the mouth of the Muskingum River. In 1808, the last buffalo thought to be killed in the state occurred near Zanesville.

Wood Bison, the type that existed in Ohio, were larger and heavier than their counterparts on the far western plains. Those in Ohio were tough, long-lived and prolific. They could run swiftly and they were adaptable to extreme heat or bitterly cold temperatures. They were ruminants who fed on grass, grass in the Ohio prairies that grew to such heights in places it concealed the barrel-chested creatures as they foraged.

Buffalos were fond of the leaves on the tall, fibrous plants known as "cane" which grew in abundance in parts of Ohio. Kentucky was especially famous for its cane fields. There were thick stands that covered acre upon acre of ground. Each autumn the buffalos migrated south from Ohio, crossing the Ohio River to the cane fields in Kentucky where the winters were somewhat less severe.

In the spring, the herds would begin migrating northward. They were masters at finding instinctively the most logical spots to cross rivers. They swam slowly and low in the water with only the top of their head and occasionally a hump showing. While it might suggest that buffalos were poor swimmers, usually they performed quite satisfactorily.

A party of hunters once watched two buffalo bulls swim across the Ohio River. Jumping into their canoe and paddling alongside, one of the men playfully grabbed a bull by its tail. Laughing and shouting, the men delighted in their aberrant ride for more than half an hour as the baffled animal attempted to dispose of his unwelcome burden.[5]

Young calves, born in the spring, kept up amazingly well with the migrating herd. One was said to have traveled over sixty miles, part of the way in deep snow—and all before it was three days old. A five month old calf was reported to have run more than fifteen miles before men, using a series of three fresh horses, finally ran it down.

If orphaned or strayed calves were not devoured by wolves, they often made the mistake of adopting any non-buffalo foster parent that happened along. Meriwether Lewis told of going for an evening stroll when a buffalo calf tread so closely on his heels that he could not elude it. Only by leaping into a canoe and pulling away from the shore did he dissuade the youngster.[6]

Hunters often encouraged the calves to follow them to camp where they were butchered and transformed into the evening meal. Indians sometimes led calves to their village by letting them suck on their fingers. When the children tired of playing with the engaging creatures, they were slaughtered for food.

A man who witnessed many such incidents expressed great astonishment that buffalo calves would stand quietly by the side of their dead mothers as the hunters who'd just killed them, skinned the carcasses. Unhesitatingly, the calves attached themselves to the men and trailed after them.

S. P. Hildreth, a doctor in Marietta, Ohio, who collected local anecdotes, compiled a book entitled *Biographical and Historical Memoirs of The Early Pioneer Settlers of Ohio With Narratives and Occurrences in 1755.* He told about a buffalo bull calf that belonged to a man in Washington County. In 1799, Mr. Ephraim Cutter, the owner, decided to derive some financial gain from the animal and exhibited it in towns throughout the East where it was a center of attraction. The bull's initial, docile temperament gradually changed and it became noticeably aggressive. One day it caught Ephraim off-guard and gored him. As a result of the injuries he sustained, Ephraim Cutter died.[7]

Buffalos have a keen sense of smell and whenever they think they have scented danger, have a strong tendency to run, often stampeding the entire herd. Edna Kenton, Simon Kenton's great niece who wrote about her uncle's exploits in *Simon Kenton, His Life and Period, 1755-1836,* related how he had climbed a tree while a herd of buffalo stampeded past him. He soon realized there was another person in a nearby tree and, uncertain if it were a friend or a foe, finally called "Come out–Show yourself." "Come out yourself," replied the other.[8] When it proved to be a man who recently had erected a cabin about forty miles from Kenton's camp, they conversed as best they could over the rumbling noise during the lengthy wait for the procession of buffalos to disappear. After the buffalos were gone, the two climbed down and each went on his way.

Daniel Boone, once caught in the midst of a buffalo herd, stood for several hours with his back against a tree listening to them thunder by him on either side. He noted that the younger cows and calves appeared to be the vanguards, followed by the yearlings and the two-year-olds. The mature cows and bulls flanked the herd and guarded the rear. Old, infirm buffalos were left to catch-up in any manner they could.

One long-time buffalo hunter contended that the simplest way to kill a buffalo was to locate the herds' watering places and shoot one of them as it drank. He reflected that a buffalo sometimes could be approached by crawling silently through a ravine or creeping in a circuitous route behind a hill. It often was surprisingly easy; but at other times, he reminisced, it required the utmost skill of even the most experienced hunter.

Buffalos served the pioneers in innumerable ways and although practically every part of the animals could be utilized in some manner, people often butchered them and saved only the steaks, roasts, and tongues. The rest of the carcass was left to decompose.

The tongue, which easily was prepared, became a favorite meal for many hunters. After carefully scorching the tongue and peeling off the skin, it was held over the embers of a fire to finish cooking.

Occasionally, there was someone who liked blanched and sliced buffalo tongue mixed with boiled udder. The brains, the stomach lining, the testicles and the unborn calves oftentimes were roasted and eaten.

The liver, sprinkled with a bit of gall, i.e., the bitter-tasting bile, was eaten warm, right on the spot where the animal was killed. Old-timers referred to the bile as "buffalo cider." They told how they would toss the livers onto burning dung and, when done, pop them into their mouth. Some Indians and buffalo hunters ate the livers raw and drank the blood, warm from the animals' veins.

Buffalos generally were in prime condition in May and June. Their flesh had a sweeter flavor then. At any time of the year, however, the meat from cows was more tender than that from bulls. Dried, or "jerked," buffalo meat was a staple food for some of the earliest hunters and traders. A buffalo hunter in the far west recalled that his meal one day had consisted of nothing but hard bread and dried buffalo meat, "an excellent thing for strengthening the teeth," he joked.[9]

In 1795, a Frenchman in the hard-pressed little village of Gallipolis, Ohio, shot into a herd of buffalos and surprised even himself by hitting and killing one. The event was such a joyous occasion that a full-scale celebration was held, with a parade, music, dancing, singing, and much wine.[10]

In the mid 1700s, a group of half-famished men, in order to stay alive, singed the hair off two buffalo hides that were several weeks old and cut them into strips which they boiled and ate.

Pioneers in a tiny settlement in Fayette County, Ohio, in 1790, were dangerously low on food when, to their immense relief, one of the men spotted a buffalo bull which he killed. They ate what they needed and preserved the rest. The hide was claimed by the man who shot it.

Buffalo hides lasted for years and wore like iron. People often slept on them for warmth. Men frequently laid one, hair side up, on the ground near their campfire and sometimes wrapped themselves in another. Daniel Constable, an Englishman, traveled over much of the land east of the Mississippi River and once wrote home to his sister that "The greatest of my misery arose from cold. I had a bag of straw, some carpeting, a large wolf skin and extra clothing, (undressing was out of the question) but with all my scheming I could not be warm, but at Cincinnati I gave $6 for a Buffaloe robe, extra large, dressed & Indian painted, this settled the matter and gave me all the comfort as to warmth from the crown of the head to the sole of my foot that I could wish for, its ample and glowing folds set wind & frost at defiance...."[11]

The Reverend James B. Finley, a famous circuit rider and prolific writer who preached among early Ohio settlers and Indians, was credited by Henry Howe, another author, for restoring him to health. Howe wrote that Reverend Finley cured his "heavy cold by the Wyandot therapy of placing him on his back on a heavy buffalo robe and toasting his feet before a fire for two days... the cure was so complete that it lasted throughout his life."[12]

Brain-tanned buffalo robes were common winter-wear for Ohio Indians in the 1700s. Many of the settlers, unaccustomed to the frigid winter weather, adopted the same attire. Settlers covered the dirt floors of their cabins with buffalo skins. Buffalo skins, or "rugs," laid over saddles eased many a long day's uncomfortable ride. Moccasins were made from buffalo hide and buffalo sinews were used to stitch them. Moccasins were stuffed with buffalo hair for insulation. So-called "buffalo socks," which were worn over moccasins or shoes for additional warmth, were a type of footwear with the hair side turned-in.

Hides from buffalos killed in the winter were covered with heavy wool and were the most sought after for making robes. Hides taken from animals killed from spring through fall did not have the heavy fur and were more useful for making shirts and leggings. The much thinner skins of buffalo calves were used for undergarments.

Softened rawhide was made into gear for horses. It was used for bags and pouches of all kinds. Books were bound in buffalo leather. Strings for fiddles were made from dried buffalo entrails. It was said that shields made from the tough buffalo bull hides could be made impervious to arrows and bullets by first burying them in the ground and stoking a well-built fire over them.

Small boats known as "buffalo boats" or "bull boats" were made from one or two bull pelts with the hair turned inward. The hides, which were stretched over willow frames and lashed together with sinew, could be made water repellent by caulking the seams with a compound of ashes and animal fat. The sturdy crafts were capable of holding up to eight hundred pounds. Larger more commodious boats might carry as many as twenty men.

Frequently, pioneer families had their own primitive tanning vats which consisted of nothing more than a hollowed-out log or a wide pit sunk in the ground. The home-tanning process was a labor-intensive operation that required several months to complete. On a much larger scale were the so-called "professional" hunters who prepared their buffalo hides to sell.

In preparation for shipment, the skins, customarily called "green hides," were rolled into bundles, each buffalo hide comprising one bundle. For ease in carrying, every bundle was wrapped in such a way that the horns protruded and served as handles. Note the illustration that appeared in the April, 1874, *American Agriculturist*, a monthly publication which featured a green hide ready for transport.

Tanneries which sprang up in small towns and in the countryside all across the state became important enterprises. Although it did not necessarily apply only to buffalo hides, an 1813 Ohio newspaper ran an ad that read, "I will give 5¢ in cash per lb. for raw hides delivered at my tanyard."[13] An 1878 paper had this item: "We noticed 100 buffalo hides lying in front of C.G. Smith's tannery and learned from the foreman that they are to be dressed for robes."[14] A clothing store in west central Ohio inserted an ad that said, "For the nicest and cheapest buffalo and fancy robes, call on S.B. Grove."[15] Not to be outdone, the same issue of the paper had an announcement from J.C. McCauly informing his customers that he had received a large stock of buffalo and wolf robes which he would sell at low prices.

Pioneer women spun buffalo hair into yarn. While it was not the finest quality yarn, it sufficed for weaving fabric to make practical winter clothing. The spun yarn sometimes was mixed with raccoon fur and used to knit socks. According to Colonel Nathan Boone, Daniel Boone's son, the socks were "quite soft and wore very well."[16]

Because yearlings and two-year-old buffalo were thought to grow the best wool, a party of five hunters once shot twenty-four of the young animals simply for the fine wool they would furnish. Even the poorest grade buffalo hair, however, was kept and woven into rope. Mixed with mud, the hair also was used to chink between the logs of the settlers' houses.

Buffalos migrated continuously. They went to rivers and streams to drink; they went to feeding grounds to graze; and periodically they traveled to salt-licks for, like most wildlife, they intuitively sensed their need for the saline substance. Salt-licks, or salt-springs, were natural deposits of the mineral where animals often congregated in large numbers. Early settlers were aware of the salt-licks and were drawn to them not only because of the possibility of finding game but also because of their own need for salt. Simon Kenton was said once to have counted 1,500 head of buffalo walking in single file order to a salt bed.[17]

Buffalos used their hooves and horns to tear up grassy areas to create dust and mud baths. The pioneers called these saucer-like depressions "wallows" and "stamping grounds" for the beasts actually did wallow, rolling first on one side and then on the other in a cloud of dust. Finished with their "bath" and kicking their hooves in the air, each animal would maneuver itself upward by thrusting its shoulders with a great heave to rest momentarily on its forelegs before springing up onto all fours.

"Stamping grounds" were made in the same manner. They were bare patches of earth where the massive creatures stood, grunting and stamping in irritation as they shook off flies. The coating of dust gave them relief from parasites and the caked mud offered protection from biting flies and mosquitoes.

It was not uncommon for buffalos to avail themselves of already-formed depressions in the ground. The book, *The Land Beyond The Mountains,* by Ray Crain, includes a map that identifies Buckeye Lake as a one-time buffalo wallow. Eons ago, the depression which became Buckeye Lake, in the corner of Fairfield, Licking, and Perry counties, was created by a glacier that slid southward across Ohio from Canada.[18]

Swamps were useful as wallows, too. The "Great Buffalo Swamp" in the southern part of present-day Licking County was an extensive tract of low, marshy land where salt rose to the surface. Buffalos gathered there in large numbers to consume the salt and to roll in the soothing damp earth.

In the spring, the normally dry wallows and stamping grounds filled with water, water so stagnant and brackish that only a buffalo would drink it. Old-timers said that was no surprise to them for they always knew that buffalos could stomach absolutely anything.

By 1867 the buffalos had been gone for over half a century but the wallows and stamps in many places still clearly were recognizable. Yet in the mid nineteen hundreds, there were reported to be remnants of a buffalo wallow north of Piqua, Ohio, a few hundred feet east of Hardin Road on the Colonel John Johnston farm.

For generations, trees in the vicinity of wallows and stamps remained mute evidence that buffalo once had been there. Entire areas, often an acre or more, were left devoid of shrubs and vegetation. Lower limbs on trees were worn away where the animals had leaned against them to scratch after emerging from their dust or mud baths. The bark on the trunks of the trees were polished smooth by the constant rubbing.

Grass eventually grew again in some of the areas— ground that farmers oftentimes wanted to put under cultivation. The son of one settler told how his father handled the situation. There was a thick alkaline layer at the bottom of the wallow on their land that made it impossible to plow so his father dumped wagonload after wagonload of sand in the bottom to help loosen the soil.[19] Everywhere the buffalos went they left trails, trails that marked their territory, crisscrossing the land in all directions. Some of the most frequently used paths were trampled by herds of the huge creatures for so many years they were two hundred feet wide and six feet deep. Indians, traders, and the earliest pioneers looked for and followed the ready-made trails, or "traces," as they often were called. The trails circled around hills and led to steep river banks made into manageable crossings by the pounding hooves of thousands of heavy buffalos. Paths through thick forests were more narrow, perhaps only eighteen inches wide where the animals had walked in single-file fashion.

Because of the advantages buffalo trails afforded, they facilitated travel and served as a prelude to our modern-day road system. It has been said that scores of twenty-first century highways and back roads originally were part of the shaggy Wood Bisons' network of winding paths.

One of the most famous buffalo traces became known as the Bullskin Trail and ran in a northerly direction across Ohio. It was used extensively by the buffalos during their annual migrations back and forth between Ohio and Kentucky. It served as an important route for the Shawnees' north-south travel and, lastly, the trace afforded easier access to the land north of the Ohio River—the land our ancestors helped settle

On February 4, 1807, the Ohio legislature decreed the Bullskin Trail the state's first highway and christened it "The Xenia State Road." From the Ohio River, it ran through Xenia, Urbana, and Bellefontaine, Ohio, and on to Detroit, Michigan. Today, that road has been renamed U.S. Rt. 68 and, more recently, sections have been designated "The Simon Kenton Memorial Highway."

Chapter 6: BUFFALO

1. Ted Belue, *The Long Hunt* (Mechanicsburg, PA: Stackpole Books, 1996), p 24.
2. Larry Barsness, "Piskiou, Vaches, Buffler, Prairie Beeves—Buffalo," *American Heritage* (Oct./Nov., 1979):p 34.
3. ——, p 33.
4. ——, p 34.
5. Walter Havighurst, *River to the West* (New York, N.Y.: Penquin Group, Inc., 1970), p 56.
6. Larry Barsness, "Piskiou, Vaches, Buffler, Prairie Beeves—Buffalo," *American Heritage* (Oct./Nov., 1979): p 25.
7. S. P. Hildreth, *Biographical and Historical Memoirs of the Early Pioneer Settlers of Ohio With Narratives of Incidents and Occurrences in 1775* (Cincinnati, Ohio: Derby & Co., 1852), pp 409-410.
8. Edna Kenton, *Simon Kenton, His Life and Period 1755-1836* (New York, N.Y.: Random House from the Doubleday edition,1930), p 70.
9. Francis Parkman, *The Oregon Trail* (Garden City, N.Y.: Doubleday & Co.,1946), p 88.
10. Larry Barsness, "Piskiou, Vaches, Buffler, Prairie Beeves—Buffalo," *American Heritage,* (Oct./Nov., 1979): pp 33.
11. J. Brian Jenkins, *Citizen Daniel (1775-1835) and the Call of America* (Hartford, CT: Aardvark Editorial Services, 2000), p 330.
12. Samuel Harden Stille, *Ohio Builds a Nation (Chicago: Arlendale Book House*, 1962), p 190.
13. (*Urbana, Ohio) Farmer's Watch-Tower,* January 13, 1813.
14. *Urbana (Ohio) Citizen and Gazette*, August 8, 1878.
15. ——, November 14, 1878
16. Edna Kenton, *Simon Kenton, His Life and Period 1755-1836* (New York, N.Y.: Random House from the Doubleday edition, 1930), pp 83-84.
17. Walter Havighurst, *River to the West* (New York, N.Y.: Penquin Group, Inc., 1948), p 56.
18. Ray Crain, *The Land Beyond The Mountains* (Urbana, Ohio: Main Graphics, 1994), p 24.
19. Larry Barsness, "Piskiou, Vaches, Buffler, Prairie Beeves—Buffalo," *American Heritage* (Harlan, Iowa: American Heritage Pub., Oct./Nov., 1979), p 28.

BOVINE

Chapter 7

The majority of the first pioneers were a poor but resolute group of people whose wants were few and who accepted hardships as a matter of course. They moved into the Ohio country sometimes walking and carrying their meager possessions on their backs or using whatever conveyances they had available. Wagons of all descriptions could be seen, usually small, relatively light-weight farm wagons. Large ones were too heavy and cumbersome for the rough, narrow trails. Rarely were the wagons new; most were old and well-used. One particularly decrepit wagon was described as an ancient assortment of boards lashed together for a wagon bed. Two-wheeled carts and even wooden sleds, referred to as "mud boats," were not uncommon. Joshua Antrim wrote that "rough sleds were the vehicles of travel."[1]

Stillman C. Larkin's *Pioneer History of Meigs County* tells about the Thurman Necox family who moved west with two yoke of oxen hitched to a mud boat. When planning their trip to Ohio in 1803, they were aware of the advantages of oxen over horses. Oxen were less expensive than either horses or mules, a critical factor for most immigrants. The multipurpose oxen were slow-moving but dependable. They served not only as beasts of burden but also as a source of meat should the necessity arise. Indians had little use for oxen and were much less likely to steal them.

The versatile Devon cattle, a breed brought to America from Devon County, England, in 1623, were the breed most generally used for oxen. These reddish-brown bullocks were sturdy and could exist on poor quality feed. They usually were docile and easily could be trained.

In spite of their hardiness, however, fatalities were high and people frequently took spare oxen. Someone in the far west once expressed amazement at the number of dead oxen he saw lining the trail. "There ought to be a Heaven for all ox that perish under the yoke where they could roam in the fields of sweet clover and timothy. . . . A traveler could find his way with no other compass or guide than his nose alone."[2]

According to some early pioneers, oxen ordinarily were named Buck and Berry so that regardless who guided them, the commands would be the same. "Gee, Buck," meant to turn right; "Haw, Berry," meant to turn left. Buck always stood on the far side; Berry always stood on the near side, i.e., the left side.

from *Historical Collection of Ohio,* Vol. 2, Henry Howe, 1904, p. 794

The Monthly Visitor, an 1872 central Ohio publication, printed a humorous story relating to the names "Buck" and "Berry." "A Davenport school master buying a team of oxen, with which to emigrate to the west, asked the dealer how he was to manage about remembering the names of so many. The dealer explained that he had only to keep in mind the names of the leaders, because if he could guide these the others would follow. The school master in starting them followed instructions very dignifiedly, to the amusement of the bystanders, by calling out, 'Whoa, Buck! Likewise, Berry, and as a necessary consequence, whoa haw the entire team!' "[3]

Because the American population primarily was composed of people who were right-handed, most individuals could control their yoke of oxen better by walking on the animals' left side. When it was necessary to move out-of-the-way, for whatever rational, drivers more accurately could see and determine the exact distance to move over. For the same reason, drivers with teams of horses usually rode astraddle the left rear horse. The modern-day system of driving on the right side of the road instead of the left as is customary in European countries has been attributed to this early practice.

Henry Munson, the eldest son of Mr. and Mrs. Ashel Munson, recounted his family's trek from Connecticut to Lake County, Ohio, with a yoke of oxen in 1821. Except for half a day when he was ill and rode in the high-wheeled cart, young Henry walked. The hazardous journey took a month and a half, the two yokes of oxen plodding methodically over the rugged trails.[4] As was customary, the boy walked along the left side of the cart to guide the animals using a six foot long pole with a leather strap—sometimes people used a bit of silk fabric—tied on one end which, when cracked, made a sharp, explosive sound much like a rifle shot. Occasionally, Henry would crack his whip and shout, "Stretch out" to urge the slow-paced beasts to forge ahead.

The big, wooden, two-wheeled carts pulled by oxen were used during the 1700s and early 1800s in large measure because the wide wheels could bridge the deep ruts in the paths. Every once in awhile, however, the grooves were too deep and too wide for even these wheels and emigrants were known to turn their largest iron kettles upside down and with their oxen drag them along the road in an effort to smooth out the roughest places.

For the better part of the year, paths were impassable for any kind of heavily loaded conveyance. Thus it was with some trepidation that a husband and wife, their four boys, and one of the grandmothers left their Erie, Pennsylvania, farm in

1815 for a one hundred and sixty mile trip to northern Ohio. Because they had been forewarned of the wretched condition of the trails, they planned carefully and arranged for the largest items to be transported to the mouth of the Huron River by boat. Traveling in "a light wagon covered with linen stretched over hoops"[5] and carrying with them only the barest of necessities, the family had high hopes of making a speedy journey. The route was even more deplorable than they had been led to expect and they were forced to stop along the way and purchase a yoke of oxen to hitch to their team of horses. With the added strength of the two powerful oxen, they still were unable to achieve more than six to ten miles a day.

Several newcomers from the state of New York, in 1799, obtained title to land in what today is Summit County, Ohio. The difficulties encountered by the group and their six oxen and two milch cows, were appalling. The men were unfamiliar with the wilderness and had no clear conception of what to expect on the way. They loaded everything they had on the backs of the oxen. "They waded fordable streams and compelled their cattle to swim those that could not be forded, passing across those streams themselves with their provisions on rafts hastily made of sticks." Their oxen were tormented with great swarms of gadflies and one of the oxen actually died as a consequence.[6]

Aaron W. Putnam, the second son of Colonel Israel Putnam, came with his father on a long, tedious journey to Marietta, Ohio, in 1788. Their oxen were "patient and steady animals who were well-suited to the difficult passes of the mountains." At one point, Aaron remembered, after they had driven the oxen onto a flatboat to cross a mile-wide river, a sudden gale raised high waves and the boat began to fill with water and sink. They quickly unyoked the bullocks so they could swim ashore. Because Aaron could not swim, he "selected one of the most active ones and grasping it by its tail with one hand and brandishing the whip in the other, directed him with his voice and occasional touches of the whip to the shore. The wind and the tide carried them one hundred yards below the landing where they reached solid earth in safety. The other ox had no encumbrance and made the land higher up. Finally, all were collected without any loss of yokes, etc."[7]

Aaron recalled another incident. To better secure their cattle from Indians, the Putnams and their neighbors at Marietta, ferried them across the Ohio River into Virginia (present-day West Virginia). On one of the crossings, the cattle all ran to the same end of the flatboat and it sank. The cattle swam to shore—to the Virginia side—but the men were left standing breast-deep in the water. People on the bank came in canoes and rescued them.[8]

Oxen could be willful. James Swisher, an Ohio native who wrote about his adventures driving oxen, said that putting a yoke on an ox always seemed to turn into a Herculean task. "That was a job I dreaded, and the more so after we had the cattle corralled in a large pen. I was nervous for some of them had horns, and, oh! what horns, nearly as long as the rest of the body; they looked frightful! Some of the cattle were wild, very wild, while others were friendly, in fact, too friendly, for they would come as far as they could get their horns through the corral fence to meet us. I do not know why, but somehow or other it became an understood thing, from the start, that I was to help yoke and tend the cattle. Each one of them was in possession of a pair of hind legs that a mule might be proud of. Probably the only thing that kept me from using my revolver was that such action might be fraught with much more danger to the persons around the corral, watching, laughing and joking. Some of these cattle were easy to yoke. We had to lasso other of them, and draw them up to a post. We would put a yoke on the one caught, and then lasso another and draw him up beside that one. We used the gentle ones for leaders."[9]

Swisher continued with his story about oxen and said, "I had never driven cattle in my life but I did not see anything to hinder me. I thought about all there was to do was to walk along and keep them in the road; if an ox shirked a little, to touch him up with the whip. But that word 'whip' brings to my memory the many painful cuttings and slashings that I inflicted upon myself. I had a whip with a lash eight feet long, near two inches in diameter at the largest part, and a stock about four feet long. This whip worried me. I could not crack it like other ox drivers did. I was continually trying. I wondered how they could make their whip crack so. I thought there must be some slight in it. My companions in the profession had been driving before; they had had experience. They tried to teach me. I would try, try again; I kept trying, and all I could accomplish would be to slash the tail end of my whip around my head and neck. I would then try the under hand lick; would succeed in cutting myself most unmercifully around the legs, or else in getting the lash all coiled and entangled about my feet, almost throwing me down. I would stop and uncoil it, and get it all straightened out, and then try to swing it over and around my head, but there was something wrong with the whip, for the snapper and my head were continually coming in contact with each other, bringing the water to my eyes. I was, of course, angry, and out of patience, but I kept my sufferings to myself. Myself and whip afforded much amusement to the boys. I was vexed to think I was outdone. I would steal out with my whip where I thought I would be unobserved, to practice striking at some object. But the boys were wide-awake fellows when there was any prospect for fun. They would steal a march on me, and lie concealed and watch me cutting and slashing away with that whip in very dead earnest. When they had laughed until their sides were sore, and their cheeks were wet with tears of amusement, they would then laugh outright, and make their presence known to me."[10]

    Wilson Vance and Philip McKinnis, of Findlay, Ohio, had their own, if somewhat different, experience with ox-whips. The two friends, each with his own cart and yoke of oxen, once or twice a year traveled to the nearest trading post of notable size. This also happened to be the home of some of Vance's relatives. When the men had completed their business, they spent the remainder of the time visiting and then headed north again toward home. Near the Hancock County line south of Findlay, the road branched into two main trails.

The trail to the right that led into town was longer by a mile or two and the trail to the left, although shorter, passed through a wide depression in the road that, year-around, was spring-fed and often knee-deep in mud. One unforgettable day when the two were returning home, Vance suggested they try the trail on the left. McKinnis favored the trail on the right. Finally, they agreed upon a unique solution. They would travel on the trail to the left; however, if either of the two carts became stuck, McKinnis had free rein to give Vance a sound beating with his ox-whip. Albeit, if both carts succeeded in driving through the mud without stalling, Vance would flog McKinnis.

Men and ox carts proceeded through the mud with no apparent effort and McKinnis was resigning himself to losing the wager when, all of a sudden, Vance's oxen sank. Gleefully, McKinnis unhitched his yoke and hooked them to Vance's and the four animals, straining together, pulled the cart to solid ground. Vance good-naturedly gave voice to a few choice expletives and, with a grimace, hunched his shoulders forward and stood as his companion flailed him with his whip. Throughout their lives, the two men remained firm friends.[11]

One of the relatives with whom Wilson Vance visited on his trips to the trading post was his brother, Joseph. Joseph, who became the tenth Governor of Ohio, from 1836 to 1838, had worked as a woodchopper at a salt works in Kentucky when he was a young boy. He had earned money enough to purchase a yoke of oxen and with his oxen delivered salt to the small settlements scattered throughout the state. In 1805, when he moved with his family to Ohio, he had continued his occupation as ox driver, occasionally making trips back to the salt works in Kentucky.[12]

Joseph's children often heard their father reminisce how he had had to unhitch the oxen and use them to test the condition of the creek and river beds before attempting to drive through with his loaded wagon. It was not unusual for him to have to unload and roll the barrels of salt through mud holes and marshy areas. Always working alone, he then would have to reload the barrels and head toward his next stop. He told of waiting two and three days, sometimes without food, before he could continue his journey.[13]

The tiny settlements and the isolated cabins scattered in the wilderness were connected only by a network of paths that led from one to the other. Regardless of the distance and the difficulty in reaching them, however, neighbor helped neighbor. Lyman Ballard of Ross township in Greene County hauled grain to a mill in Clifton for himself and his neighbors. It required a minimum of two full days. Jonah Wood, a settler who lived on the east bank of the Little Darby, near West Jefferson, Ohio, made it a practice to maintain two or three yoke of oxen in order to keep his community supplied with flour and salt. Each fall Jonah took honey, beeswax, maple syrup, skins, and feathers to Zanesville where he sold the produce and then returned to the Little Darby with the flour and salt and other necessities his friends requested.[14]

The role of salt in the lives of the early pioneers was an issue of enormous importance. Although meat could be cured by pickling or smoking, salt preserved it far more satisfactorily and it could be kept for prolonged periods of time. While many of the people resigned themselves to eating their food without the benefit of salt, it still was an essential commodity.

It was not unprecedented to have to give a cow with calf-at-side for a bushel of salt. Hunger, illness, deprivation, hostile Indians and the often bitterly cold winters, were burdens too great for some families to bear and although they had come west with courage and purpose, they gave up and went back to their former homes. One of the most difficult things to combat was the unrelenting loneliness. Women in particular were affected but there were some like Rowena Tupper of Marietta, Ohio, who said "the Ohio country [was] a 'savage land' but it "answered every expectation."[15]

Most women, nevertheless, sometimes yearned for the companionship of other women. When a family near Vermillion, Ohio, learned of the arrival of a new family only three miles away on the opposite side of the river, arrangements were made for the two women to meet and visit. On the day of the proposed visit, torrential rains had made the river impossible to cross on foot. Because there was neither a horse nor a canoe nor a raft of any kind, there seemed little chance for the women to get together until the water receded. Fortunately, the new settlers owned a yoke of oxen. The husband "fearlessly sprang on the back of one of the oxen and struggled across the raging waters" leading the other ox. The woman who eagerly was waiting, lost no time climbing onto the back of the ox the man had been leading. Turning the oxen around, they returned to the other side of the river. The women embraced warmly and talked throughout the day. At the end of the visit, the woman again climbed onto the ox's back and was escorted across the river.[16]

Occasionally, there were men who acknowledged having deep-seated feelings of isolation. Others reveled in the solitude. In 1812, a man by the name of Conner and his bachelor son enjoyed living alone in a cabin not far from today's Versailles, Ohio. The father spent his days hunting by himself in the woods for deer and any wild game he could find. He depended on his son to tend to their small crop of corn. With the aid of an old ox the son cheerfully went about his chores. Only when their supply of cornmeal ran low did the young man take a break from the daily routine. He tossed a sack of corn on the ox's back, climbed aboard, and rode the thirteen miles to the mill. Nighttime frequently overtook him. Undismayed, the son would tie the ox to a tree, build a fire, cook a meager supper, and lie down in the forest to sleep. At daybreak, he would untie the ox, load the cornmeal, climb on the ox's back and head homeward again.[17]

Riding one's oxen was not unusual. The Showalters of Harrison township in Van Wert County were one of the families who relied on their oxen for transportation as well as using them for beasts of burden. A history of the county says the Showalter sons amused themselves for hours at a time breaking their oxen to ride. They gentled the great beasts so successfully they could ride them anyplace they wished to go.[18]

Beginning in the seventeen hundreds, people of all ages and of all nationalities streamed into the Ohio territory. Single males and young families comprised the largest percentage. A great deal of the time, the sturdy, dependable oxen were at the forefront. The journeys ordinarily were well organized affairs with certain individuals guiding the oxen and, generally, boys herding the livestock. Nearly every person, old or young, had an assigned task. Feeding and caring for the oxen usually took precedence.

Whether people came in a large group or as an individual family, most tried to bring at least one milk cow with them. Twelve-year-old Thomas Oxenford led their only milch cow, Brindle, on a long rope and sternly was warned not ever to let her get away.[19]

If it were not possible to bring a cow with them, it was one of the first things a family attempted to acquire upon reaching the end of their journey. Although there rarely were many cows from which to choose, a few guidelines for making a wise selection were offered in *The Compleat Housewife, 1776.* "The first requisite is to have a good cow. One that has high hips, short forelegs and a large udder is to be preferred. The cream-colored and the mouse-colored cows generally give a large quantity and of rich quality."[20]

A singular, 19th century method of feeding cattle was offered by a Mr. Davis who observed one summer day that the tomatoes in his garden were going to waste. He resolved to see if his cow would eat them and later reported that he had "fed his cow this season, at least ten bushels of tomatoes." "The plan," he explained, "is to mix a little bran with them, (say three qts. to a half bushel of tomatoes); when fed, they cause an excellent flow of rich and delicious milk."[21]

One elderly gentleman went to great lengths to care for his favorite cow. He took her with him everywhere he went for he said her milk was the only milk he could drink or use in his food without upsetting his stomach.

A homesteader in Van Wert County, whose primary concern was to have a source of milk for his growing family, plowed corn for a neighbor at thirty-one cents a day for an entire month in order to pay for a cow.[22]

In 1797, the Entsmingers traveled with their livestock and household goods to Meigs County. The cows which they drove in front of the wagon were guarded with the utmost care. They were milked morning and evening. The milk that was not used at once was carried in canteens for consumption during the day. Cream frequently was poured in a container and placed at the back of the wagon where the perpetual joggling served as a churn to make butter. When necessary, the family stopped for a day or two in order to cook and bake and make repairs to the wagon.[23]

In 1804, Archibald McConkey moved from Kentucky to Ohio. Mrs. McConkey rode the horse and carried their baby daughter in her lap. A small son sat in back of her on the horse and in back of the little boy, rolled into a bundle, was a feather bed they steadfastly refused to leave behind. Mr. McConkey walked the entire distance leading their two cows.[24]

Jonathan Alder, the white man who was captured as a boy by Mingo Indians and taken to their village on Mad River, was adopted into the tribe. He lived with them for many years. After the Greenville Treaty he settled on land among the early pioneers in central Ohio where he kept cows, hogs, and horses. A history of Logan County mentions that Jonathan sold milk and butter to his Indian friends.

Now and again, butter served a more utilitarian purpose. Wagon wheels and axles required frequent greasing and if the customary mixture of pine tar and lard was not available, butter could be substituted. Normand MacLeod, a British army officer who assumed a post on the American frontier in the late 18th century, wrote in his journal on October 18, 1778, that he had ordered "12 Pound of Butter to Grease the Cart with."[25]

In the winter of 1830, Henry and Elizabeth Ellithorpe were married in Sandusky, Ohio. Immediately following the ceremony and braving the inclement weather they, with their six head of cattle, walked across the frozen ice on Lake Erie to their home on Kelley's Island.[26]

A young couple in Madison County began married life with nothing except an old iron cooking pot, two broken knives, two broken forks and two old pewter spoons. They did, however, own a two-year-old heifer. Money was scarce. In fact, there was little cash in circulation anywhere in early Ohio and like nearly everyone else, the young couple used their produce to barter for goods. They often remarked they had had more difficulty raising the eight cents tax assessment on that first heifer than any other tax they ever had to pay.[27]

A man and wife and their three little boys left Kentucky in 1796, and crossed the Ohio River to settle in Madison County. The first winter they spent in their new home they "had not even a sign of a bed to lie on. They had a large box, sufficiently large for him and his wife to lie in, and in the fall they gathered leaves and filled the box. They had two blankets; one of these they spread over the leaves for a sheet, and the other they used to cover with. This constituted their bed for a year or two after they came to this country. The children had to shift for themselves. In the evening, the two oldest boys would gather a large quantity of prairie hay or grass, take it into the house and pile it in a corner, and then the three little fellows would crawl under it and sleep until morning; then gather it all up and take it out and give it to the cow." The boys were responsible for feeding the cow and their mother for milking it. On cold and rainy days she was in the habit of hurrying the cow into the cabin before sitting down to do the milking, an action that earned the contempt of other pioneers in the community who considered her lazy and shiftless.[28]

Jonathan Hale and his family who settled in the Western Reserve in 1810, were prosperous immigrants and enjoyed many of the amenities of the times. Jonathan was a talented violinist. His grandchildren teased him unmercifully, nonetheless, because although his fingers were nimble enough to play the violin with consummate skill, he simply could not grasp the technique for milking a cow. When no one in the family was available to do the twice daily chore, he led the cow to a neighbor's and requested their help.[29]

The majority of early pioneers turned their cattle loose during the daytime to forage at-will in the woods. Although the animals sometimes had nothing to graze on except leaves and shrubs, it was the best most people could do.

In attempting to describe the dense forest and the uniformity of the trees in the woods where his cows roamed, a Brown County farmer explained he, an adult, easily could become confused and lose his way. Especially on cloudy, overcast days, one had to be particularly alert to his surroundings. There were many times, he said, when he could find his way home only by following the sound of the bells he had hung around his cows' necks. They always knew in which direction to go and he simply trusted their homing instincts.[30]

It was a commonplace practice to send children to find the cattle and drive them home for milking. In Clermont County, eleven-year-old Lydia and seven-year-old Matilda Osborn went into the forest to search for their cows. The girls located the cattle but when they began walking in what Lydia decided was the wrong direction, she instructed Matilda to stay where she was and then ran to try to head off the cattle. Tiring of waiting, Matilda followed the tinkling of the cowbells and arrived safely at home. Hours later, when Lydia still had not appeared, it was feared that she was lost. News of the lost child "spread rapidly through the neighborhood. Bells were rung, horns blown and guns fired, and the woods and thickets beat and scoured all night, in vain. The news flew in every direction, and a constantly increasing crowd of frontiersmen gathered in and assisted in the search." Traces of the places where the little girl had slept and had gathered berries were found. Footprints were seen and a tiny structure erected of sticks with moss covering the cracks was found near a blackberry patch. Her bonnet was recovered hanging on a bush and miles farther away an abandoned Indian camp was discovered. It was the general consensus of opinion that the Indians had taken Lydia with them. Nothing more of the child was heard.[31]

For the first pioneers, Indians were a nearly constant worry. Although many were on friendly terms; others were a threat. *The History of Adams County* recounted the experience of George Campbell. In the fall of 1792, the fourteen-year-old ran away from home because of difficulties with his stepfather. He found his way to a stockade where the inhabitants were pleased to have his help in caring for the livestock. It was his chore to herd the cows into the woods each morning and to drive them into the stockade in the evening. Late one afternoon as he was walking along behind the cattle, he spied a party of Indians who saw him at the same time and started to give chase. Level-headed George began shouting as loudly as he could. "The cattle took fright and went for the stockade on the run. The boy did the best running he ever did in his life, yelling Indian style all the time." Just as George had anticipated, men in the stockade interpreted his cries and rushed out fully armed. The Indians turned and fled.[32]

In 1794, twelve-year-old Rauie Runyan was killed by Indians. John, a distraught older brother, swore vengeance and threatened thereafter to kill every Indian he saw. Forewarned of his threat, a band of Indians decided to kill him first. Removing the bell from around the neck of the Runyan's cow and using it as a decoy, they misled the young man into thinking it was the cow he was trailing. Deep in the woods they ambushed and killed him.[33]

Stories of cattle straying too far and becoming lost were not out-of-the-ordinary. Other than milch cows which were tended-to on a daily basis, cattle and hogs were allowed to roam at-large and often were not seen for weeks and even months by their owners. In order to identify one's own livestock and lay claim to those who had wandered away, it became necessary to use special marks on their ears and record those marks with the county officials. One man listed his mark as "a crop of each ear and an upper bit in the left." Another settler used "a swallow fork in the left ear and a slit in the right."[34]

Law abiding citizens went out of their way to help each other. When notices posted on trees and public places and in early newspapers made known that livestock was missing, people who found the animals advertised the fact and the owner would claim his property. Seen in a January, 1813, newspaper was this announcement: "Taken up by William Chapman of Champaign County, German township, a Red Cow with a star in her forehead, supposed to be ten years old, a crop off the left ear and a half crop off the right ear, hipshotten in the right hip, has a young calf about three weeks old—appraised to ten dollars by Archibald M'Kinly and Abraham Magard, December 21, 1812."[35] The term "hipshotten" alluded to the fact one hip was lower than the other.

It also was common practice to announce in local newspapers that an animal had been duly identified and returned. A Hamilton County news article read: "Four head of neat cattle taken up by me sometime in January, 1801, were claimed on the tenth of Sept. ensuing by Thomas Young, living in Hamilton County, waters of Little Miami, 20th of Sept., 1802." The same paper carried the notice: "These are to certify that a cow and calf taken up by me last Feb. have been claimed by and proven to be the property of Mary Harrison, Aug. 26, 1802."[36]

Dishonest citizens sometimes paid the penalty. An 1880 paper noted that a man had been confined to a jail cell for "having mistaken the cattle of Truman Kimball for his own."[37]

Livestock running loose in the woods were at risk and oftentimes became easy prey for wolves, bears and panthers. Abraham Studebaker, for example, lost his only cow and her heifer calf. Soon after it was born, wolves killed the calf. Hoping to catch the wolves, Studebaker built a wolf trap and baited it with the remains of the calf. The wolves continued to be elusive but his own cow, "unluckily, was so overcome by curiosity as to put her head into the trap, which was sprung and broke her neck."[38]

A far less serious, actually a humorous, incident regarding cattle occurred in Champaign County. "Mr. A. McIlvaine, of Rush township, lost three fine steers a couple of weeks ago, and for several days scoured the country in search of them. He finally concluded they had been stolen and butchered, but a search among the hides purchased by the tanners here failed to furnish any evidence of it.

"One day last week, Mr. McIlvaine and a hand went into the field to see if more cattle had been stolen, and going to a straw rick noticed the nose of a steer sticking out from the straw. A clearing away of straw revealed the three lost steers. They had eaten into the straw so far on one side of the rick that the top had fallen and covered them, and they had eaten through to the other side."[39]

Mother Nature was responsible for some of the misfortunes that befell cattle. Following a severe wind storm in the mid 1800s, a cow was found wedged in the fork of a tree. Henry Howe told about an 1825 tornado in Licking County that caused great destruction. "An ox belonging to Colonel Wait Wright was carried about eighty rods and left unhurt, although surrounded by the fallen timber, so that it required several hours chopping to release him. A cow, also, was taken from the same field and carried about forty rods, and lodged in the top of a tree, which was blown down and when found was dead. . . . A yoke of oxen, belonging to William H. Cooley, were standing in the yoke in the field and after the storm were found completely enclosed and covered with fallen timber, so that they were not released until the next day, but were not essentially injured."[40]

Livestock that roamed at-large could, themselves, cause wide-scale damage. Irish-born William Lyon immigrated to Ohio where he served for a time as a chain-carrier with a surveying crew. Using his earnings, he purchased several tracts of land in southwestern Ohio and began clearing a plot. During the period it required to clear enough ground to plant a crop of corn, he slept in a hollow log. Crossing into Kentucky every week or so to restock his provisions, he was able to spend most of his time girdling the trees, cutting away the underbrush and plowing. When the soil was ready for a sizeable field, he planted his corn and waited for it to come up. Three weeks later, satisfied that the corn was doing well, Lyon went to Kentucky for much needed supplies. While he was gone, a nearby settler's cattle ambled into the field and totally destroyed it.[41]

In 1888, a cow caused a train derailment. The newspaper account read: "A cow got in front of a double header freight train. Two engines and ten cars were wrecked. Many cattle were killed. A brakeman was killed instantly; a fireman was badly scalded and another had an arm broken."[42]

Dates vary depending on the information one reads but, perhaps as early as 1790, a few enterprising cattlemen recognized the potential of the lush grasslands in Ohio—grasslands which were ideal for fattening beef cattle. The remoteness of the settlers from markets in the East resulted in such low prices for their products, these foresighted gentlemen reasoned the only recourse appeared to be a venturesome undertaking, that of feeding their cattle here and delivering them "on foot" directly to markets where there was the greatest demand—urban markets like those in Pittsburgh, Philadelphia, Lancaster, Baltimore, and sometimes New York.

The answer was "cattle drives." Although drovers had been driving herds to Cincinnati and Sandusky for a number of years, George Renick of Ross County was among the first to attempt a far more lengthy drive. In 1805 he is believed to have delivered his fattened cattle to slaughter houses in Baltimore, Maryland. Some of George's friends feared his livestock would lose so much weight on the way, he would return home penniless. George proved them wrong and soon thereafter his brother, Felix, successfully drove cattle to the East with profitable results.

A picture of Felix Renick in 1831 shows him in the process of driving a herd of over two hundred head of cattle. The drawing depicts him astride his horse and wearing a tall, black hat. He holds a whip in his right hand. A written description further tells us he is sitting on a bearskin robe laid over the saddle. He also is carrying a rifle and for protection against the sun and rain he often was known to carry an umbrella.[43]

Herds of three and four year old steers numbering from one hundred to three hundred head annually walked from Ohio to the eastern slaughter houses where they often sold from $8.00 to $11.00 per head. Unless a herd was large enough to warrant two men on horseback to control the cattle, ordinarily only the man in charge rode a horse. Hired hands walked and carried long, blacksnake whips. All were called "drovers,"— those who rode as well as those who walked.

A crew usually consisted of five drovers, including a boy who preceded the herd leading a well-trained ox. The ox performed a valuable service for cattle, like many animals, are gregarious and inclined to follow another of their species. The boy who managed the ox often carried over his shoulder a flag on a long standard to alert travelers on the road ahead of their approach. The last drover helped to keep the slower animals moving, rounded-up the strays, and arranged for farmers along the way to pasture and feed any sick or lame animals.

A cattle drive moved slowly, at the most ten miles a day and possibly only five miles if a herd of pigs were being driven behind the cattle. Turbulent rivers, ankle-deep mud, choking dust, and swarms of insects, could slow travel to a near crawl. Drovers had to depend on feeding and watering their cattle at special inns called "drove stands," where pens and/or pasture and forage could be purchased. This type of tavern and inn sprang up every few miles, especially along main roads such as the National Road. Unlike high-class taverns and inns, those planned specifically for drovers were built on the outskirts of towns.

To drive a herd on the turnpikes, drovers had to pay a toll. Rates, which were relatively uniform—from 3/4¢ to 1¢ for each head of cattle, six months old or older—were collected at the point of entry by attendants who usually were appointed by the governor of each state.

Long poles called "pikes" rested on a post at the side of the road and extended across the path. After each animal had been counted and payment made, the pike, which was arranged on a swivel, was pushed out of the way and the drive passed on to the next section of the road. Toll gates ordinarily were found about every ten miles apart. Anyone trying to by-pass a toll station was subject to a five dollar fine. Drovers generally were given permission to pay when they returned home after the completion of the drive and were given a certificate guaranteeing free passage over the other toll roads they might use. Some drovers prepaid the full amount and received a certificate for free passage. In the absence of cash, it was customary to use potatoes, apples, meat or grain.

Droves in the 1800s became so numerous a traveler once remarked that cattle, hogs, and sheep, were never out of sight on the highways. Stagecoach drivers were annoyed at the delay the long lines of livestock could cause and swore when they saw drovers' flags. By the early 1820s, more than ten thousand beef cattle were estimated to have been driven to eastern markets.

Peter Slaughter of Highland County, was one of the men who contributed to that number of cattle. It has been reported that he bought and sold over one thousand head annually and drove them to Lancaster, Pennsylvania.

Drovers returned home by the same route in order to collect the cattle that had been left behind. Because the man who directed the drive in all probability had no cash with which to pay bills as they headed to the markets, innkeepers, tavern-keepers, and drove stand proprietors generally extended credit. They, too, were paid on the return trip home.

Enormous numbers of beef cattle were required to keep armies fed during the many long years of conflict between the Indians and the white man. Herds of cattle often moved with the troops. Primarily because of the slow-moving livestock, early expeditions sometimes traveled only three miles a day.

One line of march was described in a history of Coshocton County. "Volunteers went in advance, preceded by three scouting parties, one of which kept the path, while the other two moved in a line abreast on either side to explore the woods. Under cover of the axe companies, guarded by two companies of light infantry, they cut two parallel paths, one each side of the main path, for the troops, pack-horses, and cattle that were to follow. First marched the Highlanders, in column two deep in the centre path, and in the side paths in single file abreast, the men six feet apart; and behind them the corps of reserve and the second battalion of Pennsylvania militia. Then came the officers and pack-horses, followed by the vast droves of cattle, filling the forest with their loud complainings. A company of light horse walked slowly after these, and the rear guard closed the long array. No talking was allowed, and no music cheered the way. When the order to halt passed along the line the whole were to face outward and the moment the signal of attack sounded, to form a hollow square, into the centre of which pack-horses, ammunition, and cattle were to be hurried, followed by the light horse."[44]

Captain Henry Brush of Chillicothe led a supply train during the War of 1812. Marching in single file were "seventy packhorses each carrying two hundred pounds of flour. These were followed by a drove of three hundred cattle."[45]

Selling cattle to the government became a money-making proposition for some. For others it was less than profitable. John H. Piatt of Cincinnati, Ohio, founder of the first bank west of the Alleghenies and a successful merchant, was called upon during the War of 1812 to sign a contract with the government to furnish provisions. He was given the rank of Colonel and appointed commissary general of the Northwest Army. Provisions were to include cattle. Congress adjourned before negotiations were finalized with the disastrous result that Piatt's expenditures totaled considerably more than the amount the government paid him. Appealing to his patriotism, and reassuring him he would be paid, Piatt was persuaded to continue his funding. In the end, however, the government refused him remuneration. Twice, Piatt's creditors threw him into debtors' prison. In the end, he died there "without enough money to give him a decent burial."[46]

Cattle drives, to a certain degree, were a contributing factor in an effort to improve breeding stock in Ohio. Cattlemen needed animals that would gain weight economically and easily and, at the same time, would retain their weight during the long trip to the markets. At great expense, purebred cattle were imported from England and a crossbreeding program began. When the results became evident, buyers willingly paid the high prices demanded for proven animals and their off-spring.

M. F. Rickey and James G. Gilliland walked from Van Wert County to the State Fair in Columbus, Ohio, with the intention of investing some of their hard earned money in fine stock to add to their herds. They each found a calf they wanted; one was a six month old Durham bull calf and the other was a seven month old Durham bull calf. At a time when calves ordinarily sold for $10.00 to $12.00, the two men paid $50.00 for each of the two calves. Leading their new acquisitions, the men walked back home again to Van Wert County.[47]

To own fine livestock was the goal of many an Ohio farmer. Those seriously intent upon improving their herds kept meticulous accounts of their purebred cattle and their offspring. Breed associations were established to further encourage and to keep the extensive records of genetic lines of prize winning bulls and females. The shorthorn bull, Loyal Duke of Oakland, owned by D. McMillan of Xenia, Ohio, and pictured below, is an example of the type of animal considered the ideal in the 1860s.

from *Report of The Commissioner of Agriculture For The Year 1867*

After considerable haggling, James Price, of Darke County, agreed to buy a Shorthorn heifer, the off-spring of a purebred Durham, for the enormous sum of $40.00. James' wife was astounded and said, "The beast is pretty to be sure, but there is no sense in giving half a ton of cheese for it." Several months later, Price acquired a roan bull calf for $45.00 "which astonished his wife even more."[48]

At private and public auctions in the mid eighteen hundreds, the imported animals sold for prices ranging from $400.00 to $2,225 per head—amazing amounts for the period.

In the latter part of the 19th century, a few farmers in Miami County experimented with fattening beef cattle on offal from distilleries. One operation, located across the river from a distillery in Troy, piped the waste through a four inch copper line to a pen on the opposite bank.[49]

Every now and again, articles relating to cattle were found in old newspapers, histories and documents which did not necessarily conform to this chapter but were worthy of notice. One such item was a short newspaper article captioned, "A Cow For A Baby" A man and his wife "decided to separate and the assets were divided until only the baby was left, when the father said, 'If you will leave the baby with me I will give you a good cow.' The mother considered a moment and decided that a good cow was worth $25.00 and a baby—well, pretty poor property. So she took the cow."[50]

A biographical sketch in *The History of Madison County* regarding Claudius Mitchell said he remembered how "the first pair of pants he ever wore were made by his faithful Vermont mother, who manufactured them out of hair combed from their own cow in the time of shedding in the spring, mixed and carded with common flax tow, all done by hand."[51]

The following item was an undated newspaper clipping, yellow with age, that was found between the pages in an old book. It was titled, "A Boy's First Composition—Oxen is a very slow animal, they are good to break ground up. i wood drather have horses if they didn't have kolick, which they say is wind collected in a bunch, which makes it dangerser for to keep horses than an ox. If there was no horses the people wood have to wheal thare wood on a wheelbarrow. it wood take them two or three days to wheal a cord a mile. Cows is useful to. i have herd som say that if thay had to be tother er an ox thay wood be a cow. But I think when it cum to hav their tits pulled of a cold mornin thay wood wish thay wasn't, for oxen don't generally have to raise calves. if i hed to be enny i wood drather be a heffur. but if i coodent be a heffur and hed to be both i wood be an ox. by Isaac Spiker"[52]

An 1888 central Ohio paper printed an election oriented news item: "Drawn By Four Oxen — Mr. William N. Whitely voted in characteristic style Tuesday. There always is a pleasant smack of novelty about his politics, and Tuesday this was exceptionally true. As soon as the fourth ward opened, Mr. Whitely was on hand with a tremendous wagon drawn by a team of four oxen, and containing forty-seven voters. The crowd set up a yell of enthusiasm as soon as it caught sight of the novel equipage."[53]

An 1833 Ohio temperance crusade sermon delivered by Reverend David Merrill was known as "The Ox Sermon" because it was based on several verses in the Bible. Exodus, Chapter 21, verses 28 and 29 read: "If an ox gore a man or a woman, that they die; then the ox shall be surely stoned, and his flesh shall not be eaten; but the owner of the ox shall be quit. But if the ox were wont to push with his horn in time past, and it hath been testified to his owner, and he hath not kept him in, but that he hath killed a man or a woman; the ox shall be stoned, and his owner also shall be put to death." Reverend Merrill equated the story to those persons who made and supplied intoxicating beverages and the oftentimes ruinous results. The lengthy sermon became quite famous and was widely circulated across the nation.[54]

Although the account about "The Cow That Started a War" did not occur in Ohio, it does, nonetheless, offer provocative reading. "After a party of emigrants on the Oregon trail abandoned a skinny, foot-sore, old cow along the way, a Sioux Indian found her. Needing a piece of hide to patch his worn moccasins, he killed the cow and repaired his moccasins. Someone reported the act to authorities at Fort Laramie, where, suddenly, the cow became a valuable commodity. The Indian, on learning of the situation, offered ten dollars to the owner. The owner, however, demanded twenty-five dollars which the Indian refused to pay.

"A young, inexperienced, hot-headed Lieutenant marched into the Indian camp with a detachment of soldiers. There was a dispute and gunfire was exchanged. All the soldiers were killed.

"On hearing the news, a regiment was sent to an Indian camp to punish the hostiles. When they demanded the Indians responsible for the deaths of the Lt. and his men be surrendered to them, Little Thunder, the chief, said he could not comply for his people had had nothing whatsoever to do with the tragedy. Not believing he had gone to the wrong Indian camp, the soldiers opened fire and killed eighty-six friendly Indians. Among them were women and children. Seventy others were taken captive.

"So—for the next twenty years the Sioux Indians bitterly fought the whites and the old, worn-out cow was said to have been one of the causes."[55]

This last story was found in an 1866 newspaper and was titled "Widow Jones' Cows"

"Widower Smith's wagon stopped one morning before widow Jones' door and gave the usual signal that he wanted somebody in the house, by dropping the reins and sitting double, with his elbows on his knees. Out tripped the widow, lively as a cricket, with a tremendous black ribbon on her snow white cap. Good morning, was soon said on both sides, and the widow waited for what was further to be said. 'Well, madam Jones, perhaps you don't want to sell one of your cows, no-how, for nothin', no-way, do you?'

"Well, there, Mister Smith, you could not have spoken my mind better A poor lone woman, like me, does not know what to do with so many creatures, and I should be glad to trade, if we can fix it.'

"So they adjourned to the meadow. Farmer Smith looked at Roan—then at the widow, then at Brindle—then at the widow, again,—so on through the whole forty. The same call was made every day for a week, but farmer Smith could not decide which cow he wanted. At length, on Saturday, when widow Jones was in a hurry to get through her baking for Sunday—and had ever so much to do in the house, as all farmer's wives and widows have on Saturday, she was a little impatient. Farmer Smith was irresolute as ever.

"That Downing cow is a pretty fair creature—but—he stopped to glance at the widow's face, and then walked around her—not the widow, but the cow—

"That shorthorn Durham is not a bad looking beast, but I don't know' —another look at the widow.

"That Downing cow, I knew her before the late Mr. Jones bought her.' Here he sighed, and they both looked at each other. It was a highly interesting moment.

"Old Roan is a faithful old milch, and so is Brindle—but I have known better.' A long stare succeeded this speech — the pause was getting awkward, and at last Mrs. Jones broke out; 'Law! Mr. Smith, if I'm the cow you want, do say so.'

"The intentions of the widower Smith and the widow Jones were duly published the next day as is the law and custom, and as soon as they were 'outpublished' they were married."[56]

Chapter 7: BOVINE

1. Joshua Antrim, *Historical Collections of Champaign and Logan Counties, From Their First Settlement* (Bellefontaine, Ohio: Press Printing Co., 1872), p 13.
2. Bil Gilbert, "Pioneers Made a Lasting Impression on Their Way West," *Smithsonian Magazine* (May, 1994), p 44.
3. *(Urbana, Ohio) Monthly Visitor,* December 16, 1872.
4. *Ohio Historical Review* (Columbus, Ohio: pub/ed Ralph S. Estes, n. d.).
5. Lewis Cass Aldrich, *History of Erie County, Ohio* (Syracuse, N.Y.: Mason & Co., 1889), p 510.
6. Henry Howe, *Historical Collections of Ohio,* vol. 2 (Cincinnati, Ohio: Krehbiel & Co., 1904), p 627.
7. Joseph Ware, *History of Mechanicsburg, Ohio* (Columbus, Ohio: F. J. Heer Printing Co., 1917).
8. S. P. Hildreth, *Biographical and Historical Memoirs of The Early Pioneer Settlers of Ohio With Narratives of Incidents and Occurrences in 1775* (Cincinnati, Ohio: Derby & Co., 1852), p 373.
9. ——, p 374.
10. James Swisher, *How I Know* (Cincinnati, Ohio: Press of Jones Bros., 1880), pp 207-208.
11. ——.
12. D. B. Beardsley, *History of Hancock County* (Springfield, Ohio: Public Printing, 1881), pp 85-86.
13. W. H. Beers, *History of Champaign County* (Chicago: W. H. Beers & Co., 1881), p 165.
14. Joshua Antrim, *History of Champaign and Logan Counties, From Their First Settlement* (Bellefontaine, Ohio: Press Printing Co., 1872), p 257.
15. W. H. Beers, *History of Madison County* (Chicago: W. H. Beers & Co., 1883), p 291.
16. R. Douglas Hurt, *The Ohio Frontier* (Bloomington & Indianapolis, IN: Indiana University Press, 1996), p 253.
17. Henry Howe, *Historical Collections of Ohio,* vol. 1 (Norwalk, Ohio: Laning Printing Co., 1896), pp 566-567.
18. W. H. Beers, *History of Darke County* (Chicago: W. H. Beers & Co., 1880), p 414.
19. Thaddeus Gilliland, *History of Van Wert County* (Chicago: Richland & Arnold, 1906), p 176.
20. James McCabe, *Planting The Wilderness* (Boston, MA: Lee & Shepard, 1892), p 18.

21. Julianne Belote, *Compleat American Housewife 1776* (Conrad, CA: Nitty Gritty Productions, 1974), p 10.
22. A. W. Chase, *Practical Recipes*...(Detroit, MI: F. B. Dickerson Co., 1867), p 70.
23. Thaddeus Gilliland, *History of Van Wert County* (Chicago: Richland & Arnold, 1906), p 193.
24. Stillman C. Larkin, *Pioneer History of Meigs County* (Columbus, Ohio: Berlin Printing Co., 1908), p. 164.
25. Warren Everhart, "Early Settlers" unpublished manuscript (n. p., 1957).
26. Normand MacLeod, *Detroit to Fort Sackville, 1778-1779, The Journal of Normand MacLeod* (Detroit, MI: Wayne State University Press, 1978), p 54.
27. Lewis Cass Aldrich, *History of Erie County* (Syracuse, N.Y.: Mason & Co., 1889), p 476.
28. W. H. Beer, *History of Madison County* (Chicago: W. H. Beers & Co., 1883), p. 748.
29. ——, p 290.
30. Deborah Halverson, "Journey of Jonathan Hale, *Early American Life* (Camp Hill, PA: Celtic Moon Pub., June, 1985), pp 20-76.
31. Henry Howe, *Historical Collections of Ohio*, vol.1 (Norwalk, Ohio: Laning Printing Co., 1896), p 417.
32. Nelson Evans, *History of Adams County* (West Union, Ohio: ed Emmons Stivers, 1900), p 546.
33. Tom Stafford, "Shaken and Moved," *Springfield (Ohio) Sun-News*, Aug. 27, 2001
34. W. H. Beers, *History of Madison County* (Chicago: W. H. Beers & Co.,1883), p 650.
35. *Urbana (Ohio) Farmer's Watch-Tower*, January 13, 1813.
36. Nelson Evans, *History of Adams County* (West Union, Ohio: Emmons Stivers, 1900), p 109.
37. *Urbana (Ohio) Citizen and Gazette*, January 8, 1880.
38. W. H. Beers, *History of Darke County* (Chicago: W. H. Beers & Co., 1880), p 417.
39. *Urbana (Ohio) Citizen and Gazette*, January 2, 1879.
40. Henry Howe, *Historical Collections of Ohio,* vol. 2 (Cincinnati, Ohio: Krehbiel & Co., 1904), pp 66, 67.
41. W. H. Beers, *History of Brown County* (Chicago: W. H. Beers & Co., 1883), p 372.
42. *Urbana (Ohio) Citizen and Gazette*, April 12, 1888.
43. R. Douglas Hurt, "Bettering The Beef," *Timeline* (Columbus, Ohio: Ohio Historical Society, March/April, 1993), p 25.

44. Henry Howe, *Historical Collections of Ohio,* vol. 1 (Norwalk, Ohio: Laning Printing Co., 1896), p 473.

45. Patricia Medert, *Raw Recruits and Bullish Prisoners* (Jackson, Ohio: Jackson Pub. Co., 1992), p 32.

46. Henry Howe, *Historical Collections of Ohio*, vol. 1 (Cincinnati, Ohio: Derby, Bradley & Co., 1896), pp 818-819.

47. Thaddeus Gilliland, *History of Van Wert County* (Chicago: Richland & Arnold, 1906), p 176.

48. W. H. Beers, *History of Darke County* (Chicago: W. H. Beers & Co., 1880), p 399.

49. *Miami County History,* Miami County Ohio Sesquicentennial Committee (Columbus, Ohio: W. H. Heer Printing Co.,1953), p 98.

50. *Urbana (Ohio) Citizen and Gazette*, July 24, 1879.

51. W. H. Beers, *History of Madison County* (Chicago: W. H. Beers & Co., 1883), p 747.

52. Old newspaper clipping (n. p. 1859).

53. *Urbana (Ohio) Citizen and Gazette*, November 15, 1888.

54. Rev. David Merrill, *"The Ox Sermon" (*Cincinnati, Ohio: Western Tract and Book Society,1833).

55. *Mechanicsburg (Ohio) Telegram*, May 1, 1935.

56. *Urbana (Ohio) Citizen and Gazette*, February 8, 1866.

OHIOANS AND THEIR HORSES

CHAPTER 8

Leading pack horses and mules laden with provisions and trade goods, traders and fur trappers frequently blazed their own trails as they made their way through the Ohio wilderness. These hardy and adventurous men often kept hundreds of pack animals moving between Indian camps and their own bases of operation.

Usually organized into pack "trains," each "train" traditionally consisted of six or more "strings" of horses or mules. Each "string," in turn, referred to the fourteen or sixteen unshod, invariably cantankerous, animals fastened together, one in back of the other. A rider mounted on each of the lead horses guided the way. Lead horses wore a bell strapped around their neck, useful for riders to tell how far away their companions were but easily muffled when silence seemed prudent.

Every pack animal was expected to carry approximately one hundred sixty pounds, meticulously fitted into two coarse linen sacks lashed to the packsaddle and covered for protection against rain and snow. The weight came from heavy iron ware like muskets, axes and hatchets, spearheads and knives, and agricultural tools, much of which was used to barter for the Indians' furs.

In the 1740s, Irish-born George Croghan, a licensed English trader who was noted for his tact and fair dealings with the Indians, oftentimes hired more than fifty men with their packhorses. From his center of business in west central Ohio, at Pickawillany, he sent his men in all directions not only throughout Ohio but also far south of Louisville, Kentucky, and into the Carolinas where he traded with the Cherokee Indians. Everywhere he went he extolled the wonders of the Ohio region and expressed amazement at the great variety of game he saw there.

At one point, much to his consternation, all of Croghan's pack horses were seized in order to obtain additional transportation for military supplies during one of the country's long campaigns against the Indians.

The safety of packhorses during a military engagement was of the utmost importance because a large percentage of the loads was composed of ammunition. In a 1762 movement against Ohio Indians, officers issued strict orders to hurry the pack animals into the middle of the detachment at the first sign of an attack.

Packhorses and their drivers played a vital role in the War of 1812. The horses carried not only ammunition but also supplies of all kinds for the men, including whiskey. While the Northwest army rendezvoused in Urbana, Ohio, the following notice appeared in a January 13, 1813, *Farmer's Watch-Tower*, the town's first newspaper.

"General Orders
Chilicothe, Nov. 25, 1812
To facilitate the movements of the north-weſtern army, it is determined, that every ſodlier or non-commiſſioned officer of the militia of the ſtate of Ohio, who ſhall engage and ſerve the United States as waggon or pack-horſe driver, for ſuch time as the United States' quarter-maſter, or commiſſary, ſhall preſcribe; ſhall each have credit on the books of their reſpective companies for a complete tour of duty.

R. J. Meigs
Chilicothe, Nov. 25, 1812"

(Spelling and punctuation are copied exactly as it appeared in the paper.)
(R. J. Meigs was Return Jonathan Meigs, Ohio's governor)

Enthusiastic individuals, many years prior to the War of 1812, had begun wending their way into Ohio, often utilizing pack animals to carry their belongings. In the first five years following the Greenville Treaty, which was signed in 1795, it was estimated that approximately thirty thousand people came to the territory.

In 1796, a group of settlers heading north from Cincinnati led packhorses so encumbered with agricultural and household equipment they were forced to a snail-like pace. Children too small to walk were placed in creels secured to the sides of the horses. "It was a difficult matter to ford the creeks without getting the freight and the women and children wet. Trees were cut down to build footbridges across the smaller streams. Rafts were constructed to carry the contents of the creels and the women and children over large creeks while the horses and cattle swam," wrote one of the participants.[1] The creels of which the writer spoke were hampers woven from hickory shoots.

Creels were a universal way of conveying baggage and it was not unusual for small children to ride in them. Moses Dooley brought his family to present-day Preble County in western Ohio in 1805. With everything stowed on packhorses, parents and older children walked. The baby sometimes was held in its mother's arms; sometimes tucked into the center of one of the creels. The procession moved slowly and rested often.[2]

A former school teacher from the East who accompanied a party of emigrants, wrote that when everybody was assembled for the trip, he noted that a few of the people were walking and others were on horseback. He laughed out loud when he observed that in one of the creels was a featherbed and from each side of it " a young head poked out, feathers in their hair but with big grins of delight. . . ."[3]

At night, the horses were hobbled so they could graze but not wander away. Bells were hung around the horses' necks in order to locate them in the morning by listening for the rhythmic tinkling sound. Still, it sometimes necessitated hours of wandering before one would hear the bells. Searchers frequently grumbled that by the time they finally located the horses their clothing was soaking wet from walking through the dew covered weeds that often grew as high as their head.

Sometime prior to 1820, John Towel, his wife, and two children, came to Ohio from Frederick County, Virginia. Most of the way they walked and led the horses. Except for two featherbeds, they had little luggage—a few cooking utensils and their clothing. They'd been on the trail only a couple of days when, during the night, one of the horses nearly tore off a hoof. For an entire month they were compelled to make camp while waiting for it to heal well enough to travel again.[4]

For those who owned them, wagons were a decided advantage for immigrants could bring larger and more unwieldy items. Accidents, however, were common. Wagons stuck in the mud, horses broke loose, harnesses gave way and axles had to be repaired.

Jonathan Hale felt he had made excellent time with his wagon and team of horses traveling six hundred and forty six miles from Glastonbury, Connecticut, to the Western Reserve in northern Ohio—all in twenty-eight days and through sometimes rough terrain.

"One day he ventured to sport for a few pigeons, loading his gun while leaving his team untied. The ensuing shot frightened his horses to such a degree that they started into a gallop and from a gallop into a full run which soon upset his wagon casting both of the horses upon their beams ends. The wagon's whiffletree hooks were broken and, as they learned later, so were the family dishes. . . . Because of his experience with the dishes he wrote to his wife in Connecticut to stow a set of tea cups and some plates in a tight box among her clothing. Other items not available in Ohio were needed and she was to bring cloth, tea, gunpowder (to sell), leather, pepper and spices, drugs, and medicines, a copper kettle and a plowshare, and nails. He also advised her not to bring any more heavy articles than she was obliged to. He said to bring two pair of shoes apiece."[5]

When the Frederick Bonners left Petersburg, Virginia, on April 1, 1803, their four-horse teams pulled two high-wheeled freight wagons. They had everything they considered necessary for establishing a farm and a home in a new land. A two-horse hitch pulled a smaller wagon loaded with food, clothing and tents, supplies needed while enroute. Following the caravan, the children herded horses, cows, sheep, and geese. Averaging eight to twelve miles per day, the trip to Xenia, Ohio, took them nearly two months.[6]

From their former location in Albany, New York, the Morgan family spent three months reaching their destination in Ohio. An eight-year-old daughter rode one of the horses and guided the lead team. Years later she recalled how the road had been little more than a trail through the woods with just enough underbrush cut away to allow the wagon to pass.[7]
On the best of days they averaged thirteen miles a day.

Because attacks from bands of outlaws were not uncommon, someone generally stood guard at night while enroute to their new homes. W. Willshire Riley of Van Wert County wrote about his family's experience with renegades. ". . . the Cumberland, or National Road, was being built in different sections, and large gangs of Irish laborers with some negroes were at work. These men often committed outrages on travelers by felling trees across the road, and demanding pay for their removal. They tried the game on father, but as he was a large and powerful man, well armed and resolute, he soon taught them better manners, and we were suffered to pass, where others had been forced to pay these highwaymen . . . . Near the top of Laurel Hill we passed a new grave, surrounded with new pickets made out of oak, said to be the grave of a traveler murdered for his horse and money but a few days before."[8]

As readers may recall, Jonathan Alder, although captured and reared by Mingo Indians, eventually was reunited with his biological family. After remaining with them in Virginia for two years, he persuaded his widowed mother, his new wife, her family, and several of his brothers and their families to relocate in Ohio. The latter part of August, in 1806, acting as guide, Jonathan led the assemblage to central Ohio—the very heart of the wilderness to which he had grown accustomed and preferred to all other places. In preparation for the lengthy trip, they "bought one large wagon and harnessed six fine horses to it and started out." They took the most direct road from Wythe County, Virginia, to Gallipolis. "At a great many places we locked both hind wheels of our wagon and then cut down a tree and chained a log to the rear axletree of the wagon to act as a brake because the hills were so steep and long. We ferried the Ohio River and then took the road for Chillicothe and from there the road for Sandusky, passing up the Scioto to the mouth of the Big Darby and up that to our landing place for the house that John Moore and I had built a few years before."[9] In his absence, the land had been sold to someone else so Jonathan traveled to Franklinton (today's Columbus, Ohio) to see his good friend, the surveyor and founder of Franklinton, Lucas Sullivant, who sold him a tract of land.

Robert Porter and family joined the tide of immigration and started toward Ohio with a wagon and a team of horses. One of the horses died. The wagon was too heavily loaded for one horse to pull so they sold the wagon and disposed of some of their belongings. Seating Mrs. Porter and their infant son on the horse, they packed what they could in a bundle, tied it in back of the saddle and confidently continued their journey.[10]

A young couple from North Carolina looked forward to settling along the Ohio frontier. In addition to a horse, they owned a small cart. With everything piled into the cart there was little room left for the wife to ride, so holding the baby in her lap, she sat in a rocking chair tied to the top of the load.[11]

John Christian Norman and his wife, Mary Magdalene (Polly, for short), left the foothills of the Blue Ridge Mountains in 1805 and traveled alone over the mountains to Ohio. With only one horse, John Christian walked and Polly, who rode the horse, carried their infant daughter, Savilla. Their meager belongings easily fit into the saddlebags. Mr. Norman, already a noted wagon-maker in Virginia, resumed his trade in Ohio and the family prospered. The Norman's six room, brick house, begun in 1822, was one of the first, if not the first, fine houses to be constructed in the community.[12]

George Bixler and his brother-in-law had their sights set on living in Ohio and the two men journeyed ahead of their families to locate suitable land. For over five hundred miles, with one horse, they took turns riding while the other walked.[13]

Mary Christie and Louis Bancroft were married in Clark County on April 8, 1819. Their honeymoon consisted of a horseback ride, both mounted on the same horse, from the bride's home to their new cabin in the-then-small-village of Springfield, Ohio.[14]

Nathan Cory joined a party of emigrants, all of whom were intent upon finding a piece of land they could call their own in the newly opened Ohio territory. Years later, Nathan wrote about arriving in Fayette County in 1790. "His effects," he said "consisted of a wife and six children, a pony, and a dollar in cash. The money was spent for a pack of salt."[15]

In 1803, with little money but much tenacity, Edward Pence settled his growing family on a one hundred and ninety acre farm in what now is Jefferson township in Adams County, Ohio. Only by trading his horses, which were worth from $35.00 to $50.00 each, was he able to generate enough capital to purchase the property.[16]

In 1810, an old black mare was exchanged for a twenty acre parcel of land in Perry County. That twenty acres is the present-day site of Thornville, Ohio.[17]

In 1811, Isaac Gray exchanged his whole farm in Virginia for a wagon and two horses. Looking forward to a more prosperous future, the Grays packed all their possessions and, with some apprehensiveness, started for what today is Wayne township in Champaign County, Ohio.[18]

A man who came to Madison County drove a team of horses pulling a wagon loaded not only with essential supplies but also with a large box of clocks. The clocks, which sold at a great profit, were a rarity in rural Ohio for nearly everyone kept track of the time of day by watching the course of the sun. Unfortunately, the bank with which he conducted business went into receivership, leaving him penniless. His creditors attached his chattel property but when even this proved insufficient, he learned he was to be thrown into debtor's prison. With $1.30 in his pocket, all the cash the family had, the disheartened fellow accompanied the sheriff and they started for town. Along the way he remembered that if a debtor had no money, the creditor was responsible for his maintenance while he was in jail. He secretly buried his coins under a fence rail and proceeded into town with the sheriff. When he was searched and found to have no money, his creditor refused to give bond for the prisoner's care and the man was set free. On his way home, he dug up the $1.30.[19]

It wasn't easy to acquire a horse and a wagon. When a horse generally cost $50.00 and a wagon nearly as much, settlers without capital often worked for wages two or three years before being able to afford both a wagon and a team. Those who could manage both, however, usually fared better than the average man. Anybody with a wagon and a sound team of horses was in almost constant and could hire-out as a teamster for the amazing sum of $1.00 a day.

Measured by early nineteenth century standards, Isaac Blanchard was a wealthy man for he owned fifteen wagons, each of which was pulled by eight-horse hitches. For almost two years during the War of 1812, he held a contract to haul military provisions and ammunition. Floated down the Ohio River, the heavy cargo was transferred to Blanchard's wagons and then taken northward on the Bullskin Trace, today's U.S. Rt. 68, to Sandusky, Ohio.[20]

A Miami County pioneer who kept a daily account of his family's trip was unable to locate the horses one morning. After listening in vain for their bells, he concentrated on looking for their tracks and eventually realized they were heading back in the direction of the family's previous home. He retraced his steps to the campsite where he explained the dilemma to his wife and children. Without the team, the wagon was of little use and they basically were stranded, so leaving the family and the wagon where they were, he trailed the horses back over the Allegheny mountains to their old home in Maryland. He obtained necessary supplies, mounted one of the horses and, leading the second one, rode into camp several weeks later where the family was waiting.[21]

Many old journals and diaries tell about the circumstances of poverty-stricken settlers who came to Ohio with neither horses nor cattle, often walking the entire distance—people like the Moore's in Fayette County. They arrived with little to their name but youth and determination. Life was good but after a few years Mrs. Moore longed to see her family in Kentucky. Her husband made arrangements for her to use a neighbor's horse with the only stipulation she have the animal shod and that she bring the owner a quart of apple seeds. Both requests were fulfilled and from that seed the first apple orchard in southern Fayette County is said to have been established.[22]

Abject poverty was not necessarily the norm for everyone, however. There were those who came with wagonloads of fine household furnishings and, furthermore, were prepared to buy some of the luxuries of life, even at frontier prices. As early as 1793, some of the more prosperous people willingly purchased imported wines in places like Cincinnati. Merchants there stocked silver, bone china, silk stockings, Morocco and kid shoes, parasols and elegant wallpaper—status symbols in the rugged new land.

Thomas Worthington, for example, was a surveyor by profession and an affluent gentleman from Charles Town, Virginia, (now a part of West Virginia) who was conducting business in Ohio. Following two days of heavy rain, he decided he would wait no longer but proceed with his work. While attempting to ford a creek in Ross County, both the horse he was riding and the packhorses he was leading lost their footing on the slippery rocks and were swept sixty yards downstream before they could struggle ashore.[23] Mishaps such as these, Thomas once remarked, were more typical of travel in Ohio's earliest days than they were unique.

During the months Worthington spent surveying in Ross County, he had occasion to locate land that admirably fit his own requirements and on Wednesday, March 14, 1798, he and his family, with all their possessions, left their comfortable home in Virginia for a new farm near Chillicothe. Thomas Worthington, who was to become Ohio's sixth governor, and Eleanor, his wife, sent most of their belongings by flatboat. They came well prepared to lead as gracious a life as possible in the deep forests of a strange country. They "collected their furniture, including two lovely pier glasses inherited by Eleanor from her mother, and in addition, had gathered together the family silver and linen, farming implements, pots and pans, chickens, fruit trees, shrubbery, and seeds of every kind."[24] They also brought several servants.

Thomas Hulme, a friend, visited with the Worthington family on their eight hundred acre farm, Adena, nearly a quarter of a century later. He was impressed with the Worthington's farming methods and with their beautiful house that had been designed by the prestigious Benjamin Henry Latrobe. Above all else, however, Hulme remembered the tremendous manure pile "growing out of and surrounding his barn that was larger than the barn itself." "Worthington," said Hulme, "was threatening to move his barn to get away from it."[25]

Drawn by Henry Howe in 1846.

ADENA.

Some of the Ralph Hunt family moved from New Jersey to present-day Clark County in 1823. Considered people-of-means, they purchased land overlooking the Mad River Valley, and built an imposing brick house. In 1828, the other members of the family arrived, bringing their goods in covered wagons. Carefully wrapped in thick comforters and resting across her knees, Lydia Hunt, the mother, carried a delicate old English mirror that had been part of her dowry.

In anticipation of raising fine saddle horses, they brought with them a famous stallion they had used with much success in New Jersey. The Hunts stood their stallion at stud and they and their twelve children, for decades, raised horses that were the envy of the countryside. They also operated a large dairy farm.[26]

Found among the Hunt family papers was a humorous hand-written invitation for a horse's funeral: "The funeral of Grey Kate Gillett will take place (corpse and weather favorable) Sunday, Feb. 29th between the hours of 6 a.m. and 7 p.m. from her late pasture in Moorefield. Services by a one-horse chaplain. Pallbearers Duke and Diamond. The friends of the family are invited to come and bring spades, to come after breakfast, bring their dinner and go home before supper. No wake will be held or any whisky allowed in pasture. 'Nelly Gray' will be sung by the choir of Old School Presbyterian 40 grave (sic) Colonial Church, at the moment of chucking her in, after which Te Deum will be sung by any Christians that may accidentally be present. The whole to conclude with a quarter race led by the 'Dun Horse.' "[27]

Elizabeth Hughes, daughter of a highly regarded and relatively well-to-do Virginia minister, kept a diary and in it recorded many observations about their move to Ohio. She wrote that her parents and the six children rode in a two-horse carriage and, in order to oversee their progress, followed in the wake of their two wagonloads of fine furnishings. Each of the two wagons, Elizabeth said, was drawn by four-horse hitches. The trip was long and tiring but her entries reveal little of their discomfort. Mud and overflowing streams and rivers appeared to have been their greatest obstacles.[28]

Steep banks, sudden changes in riverbeds, and unexpected holes in rivers and streams always were cause for concern. Even waterways on frequently traveled paths were prone to constant transfigurations following hard rains and flooding. One party of immigrants, between Columbus and Springfield, Ohio, found it necessary to unload their wagons and transport all the goods on horseback across the Big Darby. Trip after trip was required. Finally, with everyone's belongings stacked on dry ground on the west bank, the empty wagons were dragged across one at a time by means of ropes tied to the horses. A few of the men and older boys assumed the perilous risk to themselves and swam alongside each wagon to prevent its capsizing. Nearly an entire day was spent completing the operation.[29]

In 1803, three families joined forces and journeyed together to Greene County, Ohio. Each of their two tarpaulin-covered wagons was drawn by two teams of horses. A one-horse wagon carried their everyday provisions and offered respite for the women when they grew tired of walking. The trip essentially was uneventful until they reached the Little Miami River. The wagons with the four-horse hitches crossed easily. The single horse that was pulling the light-weight wagon, however, stumbled in the middle of the river and was unable to regain its feet. Men hastily reached the stranded horse and by means of a line secured to the other horses contrived to pull both horse and wagon to safety.[30]

There are records of early pioneers who, in order to reach their destination in a more expeditious manner, arranged for their effects to be sent down the Ohio River on flatboats while they themselves traveled overland with the wagons and teams of horses. Old manifests indicate that as many as forty people in addition to tons of freight could be conveyed on flatboats. The cargo, which ordinarily arrived either at Marietta or Cincinnati ahead of the wagons and teams, was stored in warehouses until claimed by the owners. Loading their possessions in their wagons, emigrants would continue the trip.

In the spring of 1792, the William Harmer family floated down the Ohio River in their flatboat. At Cincinnati, instead of selling it as many people did, they carefully dismantled the boat. Loading their goods onto the wagons, including the lumber from the flatboat, they bounced over rugged trails to Montgomery County where they purchased a section of land in Mad River township and constructed their house with the lumber from the boat.[31]

In 1800, a company of settlers immigrating to Washington township in Montgomery County were rowed down the Ohio River in pirogues, hollowed log, canoe-like boats which could carry up to four ton. Paddled by four, robust men, the long boats, filled with women, children, provisions, harness and wagons, could travel approximately twenty to twenty-five miles a day. The horses were not taken aboard but were brought overland by the older boys under the supervision of one of the men. In Cincinnati, men and horses re-joined the families and they finished the journey together. The only major accident took place when one of the pirogues overturned and a feather tick with a baby wrapped in it floated on down the river. The baby was saved, no worse for wear, and the feather tick, although damp, was as good as new.[32]

Matthew Hodge, who traveled alone from Virginia, selected a tract of land he wished to purchase in Clark County. He returned to Virginia, put his money for the down payment in a wallet and placed the wallet in his saddlebag. "When he reached the Ohio River, it was in a swollen condition and when Mr. Hodge attempted to swim his horse across, the rapidly flowing current overturned the saddlebag and the heavy wallet went to the bottom. Many experiences had before this tested the traveler's bravery and ingenuity, and without the loss of a moment he was equal to this call on his powers, and diving to the bottom he secured the precious purse which represented, probably, the savings of years."[33]

After the National Road reached Ohio in 1825 and was constructed as far as Springfield by 1837, traffic from the east increased dramatically. One observer remarked, "The wagons were so numerous that the leaders of one team had their noses in the trough at the end of the next wagon ahead."[34]

Stagecoaches, too, began traversing the National Road and it has been said it was not uncommon to see from eight to ten four-horse coaches fairly bursting with passengers and baggage at the Red Brick Tavern in Lafayette, Ohio, a well-known stop on the heavily traveled road. It was reported that sixteen stagecoaches often could be seen, each way, every day.

Strong, fast horses were mandatory for the stagecoach trade. Speed, on the other hand, was not the primary criteria for most early Ohioans. Because the vast majority of the land was covered with dense forests and crisscrossed with streams and rivers, sturdy, rugged horses were a greater asset than fast ones.

As a consequence, one of the most frequently asked questions when trading for a horse was, "Is he a good swimmer?" People on horseback valued a horse that was vigorous enough to swim across a river with the rider still in the saddle. Someone who had had the experience described the procedure as follows:

"On entering the water we raised our feet to the saddle behind us, and thus kneeling as it were on horseback, passed dry-shod through the swift current."[35]

Reverend Elnathan Gavitt was thankful to own a hardy swimmer when he crossed the Auglaize River. "During the spring and fall when too full to ford, I would take off my clothes and tie them up, with my saddle bags, on the pommel of my saddle to keep them dry. Turning my horse into the stream, I would take hold of his tail, and when I could no longer touch bottom with my feet, I would float upon the surface until I reached the opposite shore and, putting on my clothes, start on my journey. Sometimes I would do this two or three times a day, in coldest weather of winter."[36]

In 1819, Reverend Raper, a minister whose circuit included Allen County, was swimming his horse over a swollen creek when both he and the horse nearly drowned after the horse became entangled in a submerged tree limb. With painstaking effort Reverend Raper was able to grasp a branch hanging overhead. Holding his head above water, he rested for a bit then, hand-over-hand, pulled himself to the bank. The horse, meanwhile, had kicked free of the tree limb and struggled to firm ground where it stood waiting. The minister, soaking wet, climbed on the horse and they continued on their way.[37]

Not all horses could be coaxed to enter water—Peter Clyde's for example. Clyde, a lay minister, was accustomed to walking to a site on the north bank of the Little Miami River in Greene County to preach. He did not own a horse but one Sunday was emboldened to borrow one from a neighbor. Although not totally unfamiliar with horses, neither was he overly comfortable with them. All went well, nonetheless, until he reached Burn Creek at Oldtown where the horse refused to cross the creek. Try as he would, Clyde could not make the animal step into the water. Finally, in disgust, he turned the horse around and rode back to Xenia. He returned the horse to its owner and walked the twelve miles to the Little Miami River, preached the sermon and, at its conclusion, walked home again.[38]

As a rule, horses were intelligent animals and capable of displaying sound judgment. They could be trained to obey commands and ordinarily were eager to please their handlers. Horses did need extensive grooming, however. Ohio histories tell how immigrants who came with horses had to stop for the night well before twilight in order to un-harness their teams and check them for sores that the constant rubbing of the harness could cause. Horses needed to be rubbed down and have their feet examined and cleaned. They had to be fed and watered. They could be finicky eaters and people often had to carry corn and cracked oats in their wagons if no other suitable forage was close at hand.

"At sunset their days journey finished...the horses are unharnessed, watered and secured with their heads to the trough or else hobbled out to grass."
from Henry Howe's *Historical Collections of Ohio*, 1851

Horses, providing they were allowed to rest periodically, and were fed adequately, were capable of working hard in the fields. Agile horses that could roll over on their backs to scratch themselves or simply to frolic, were said to be "worth their weight in gold." Ben Riker as a boy in St. Paris, Ohio, said his father told him, tongue in cheek, that if a horse could roll completely over it was proof of pure lineage.[3]

Occasionally, horses succumbed to epidemics that swept the region. There was a newspaper report about horses in the Zanesville area having diphtheria.[40] In 1819, an epidemic of "sore tongue," thought to be caused by dry weather or by a poisonous insect, claimed great numbers of livestock in Ohio. The remedy was an involved and lengthy one but horses were such an expensive investment, families did everything within their power to save their animals.

Initially, horses were of less importance to Ohio pioneers than oxen because the need for clearing the land and hauling the immense virgin timber required the strength and stamina of oxen. Although attorneys, physicians, and most itinerant preachers customarily made their way on horseback, and fur trappers and traders needed them for their line of work, it was not until conditions were more favorable that horses became the usual mode of transportation for the general populous.

Depending on the ground underfoot, a person on horseback could average twenty to fifty miles per day. Pulling a wagonload of goods, horses might average thirteen miles a day as opposed to a yoke of oxen's slow, ponderous six to eight miles a day. Horses were unexcelled in carrying a rider comfortably and swiftly on a long journey. Men familiar with the cavalry technique of riding a horse hard and fast could cover many miles in a day's time. They knew to ride for no longer than an hour at a fast trot, then halt, unsaddle the horse and allow it to roll, graze, and relax for fifteen minutes. There were people, on the other hand, who carried another old cavalry axiom a bit too far—"a horse can go wherever a man can travel."[41]

Most people liked horses. They enjoyed working with them—raising them, riding them, driving them, training them, racing them, and trading them.

A gentleman in Logan County was not especially concerned about his horses' pedigrees but just wanted "good" ones. This same gentleman offered the information that in his vicinity "the early native stock of horses was known as 'Virginia Spot' because some of the animals were brought from Virginia and were a vivid calico color."[42]

There were men like John S. Rarey, a Franklin County native who made a career of taming and training horses. He was one of those people whose calm touch and understanding of horses stood him well for he was renowned not only throughout the United States but also in Europe. 1897 literature listed him as a "celebrity" because of his ability to manage the fiercest of horses. It was in England that he gentled, in only one day, "the most vicious stallion in the country", a ferocious racing colt by the name of Cruiser. Rarey's book, *Treatise on Horse Taming*, sold well and was translated into several foreign languages.[43]

Men other than John Rarey had the ability to train horses by applying patience and fortitude. There was a long article in a newspaper about a horse that was disposed to travel for a short distance then stop suddenly and refuse to move any farther. A gentleman who had acquired the well-bred animal for far less than its market value because it was a notorious "balker," felt he could cure the horse.

Choosing a pleasantly cool day for the first trial, he harnessed his new purchase, hitched him to a wagon, picked up the reins and they started off. Approximately a mile down the road the horse stopped, looked around at the driver in the wagon and refused to go any farther. Unperturbed, the man sat quietly and waited, occasionally flicking the reins for the horse to move on. As often, the horse steadfastly remained where he was. For more than two hours man and horse played a waiting game, the horse now and again looking back over his shoulder at the man. Eventually, the horse responded and started down the road again. The man drove little more than a mile farther before turning around and heading for home. He unhitched the horse, patted him, and with never an unkind word or deed, led him to his stall in the barn.

The next day, the same thing occurred, except that the horse traveled a bit farther and did not hesitate quite so long as the day before. Day three was a repeat of the two previous days; however, the horse balked but a short time. The lesson was repeated on several occasions until the horse readily responded to the driver's commands. Asked to comment on the manner in which he had broken the horse of its bad habit, the gentleman responded, "The fact is, it was a test of will for the horse and for my patience as the driver. The animal broke his own will and cured himself."[44]

People had different criteria for choosing a horse. Most were quite aware of temperament and conformation while others were more interested in pedigrees and in performance. Some took under advisement information printed in an 1830 book, *The Practical Horse Farrier,* which offered suggestions for "A Good Riding Horse." It said: "His eyes should be a lively yellow or hazel. . . . A dark bay is the best color for a horse; although there are other colors which are by no means to be despised."[45]

Age usually was of significance when purchasing a horse. Pasted on a blank page near the back cover of the 1830 book was a newspaper clipping that presented rather a questionable system for determining the age of a horse. It read in part: "After the horse is nine years old, a wrinkle comes on the eyelid at the upper corner of the lower lid, and every year thereafter he has one well defined wrinkle for each year over nine. If, for instance, the horse has three wrinkles, he is twelve; if four, he is thirteen. Add the number of wrinkles to nine, and you will always get it."

People preferred sound horses but didn't always get them. Argus Cowan of Mechanicsburg traded horses one day. "After the saddles and bridles had been transferred the man said, 'Now, Mr. Cowan, we have traded, tell me what faults this horse has, if any.' 'Well, I will. He has two bad faults. He is very hard to catch, and he is of no account when he is caught.' "[46]

Henry Howe, author of several volumes of Ohio history, traveled extensively gathering material not only through Ohio but other states as well. On one of his fact-finding trips, he purchased a horse "warranted to be sound." Nonetheless, the horse turned out to be "a regular pounding machine; it took 50 miles of riding to get there, 25 by the road and 25 up and down in the air." As Howe left the inn where he'd spent the first night following the purchase of the horse, the innkeeper said, "Mr. Howe, did you know your horse was blind of one eye?" They went to the stable and looked at the eye. "It did look queer, milky-white spots were in it. 'Now wave your hand beside it,' the innkeeper said. I did so and it didn't blink a bit. It wouldn't have blinked in a gale of wind blowing 100 miles an hour. I rode him back 25 miles by road and 50 miles up and down in the air; at least so it seemed to me, so sore had I become from his dreadful pounding."

Howe returned the beast to his former owner who denied the horse was blind, saying " 'No, he is not blind, Mr. H., he is only a leetle wake in one eye.' Nothing was left me but to walk and I did walk before I got through in my successive trips more than a thousand miles."[47]

And then there were people who may have taken to heart a rhyme about selecting a horse.

> If you have a horse with four white legs,
>   Keep him not a day.
> If you have a horse with three white legs,
>   Send him far away.
> If you have a horse with two white legs,
>   Sell him to a friend.
> If you have a horse with one white leg,
>   Keep him to the end.
>
> author unknown[48]

In 1840 there was a certain doctor who derived great pride in the splendid horses he rode. He took careful stock of other horses he saw as he called on patients in the country for he felt there always was a possibility of trading for an even better one. In the course of time he became engaged to a young woman in the neighborhood but it wasn't long before he was accused of showing far more interest in one of his prospective father-in-law's horses than he was in his bride-to-be. The newspaper article read:

"A country doctor, of homely breeding, courted a brisk girl, the daughter of a farmer, who was persuaded to marry him, he having a pretty good estate. Accordingly the day was appointed. But shortly after, spying a mare on which the old man used to ride, and which for her easy gait was much esteemed, he, the doctor, desired to have her given in to complete his matrimonial bargain, but being refused, he flung away in a huff, and told the father he might keep his daughter. The girl was delighted with this rupture; but soon after the doctor repented of his folly, and came again to see her, when she was at home alone. She pretended to have no knowledge of him, 'Why, it is strange,' said he, 'that you should so soon forget me, I am your old admirer, the doctor.' 'I cry your mercy, sir,' replied she, 'I do remember me of such a person; you are the gentleman who came wooing my father's gray mare. Your mistress is grazing in the orchard, and you may make your addresses to her if you please.'"[49]

M. W. Barger also enjoyed fine horses and he attended to them with meticulous care. It's just that he sometimes was a bit absent-minded. A typical example was one Sunday morning when he forgot that he drove to church and rode home with a friend, leaving his horse tied to the hitching rack at the church.[50]

Andrew Sloan was a man who took great pleasure in handling horses and was anything but absent-minded. At the age of twenty-one when he moved to Zanesfield, Ohio, he worked as a harness maker and, on the side, began buying and selling horses. During the Civil War he acquired horses for the Union

Army. Always interested in giving his animals the best care possible, he, or perhaps his son, Earl, formulated a liniment to use on lame horses. Shortly thereafter, Andrew dubbed himself "Doctor" and advertised his services. It was a successful endeavor. Especially successful was the sale of the liniment which sold under the name "Sloan's Liniment." Comprised primarily of turpentine oil with parts of pine oil, an herb of the nightshade family (capsicum), and sassafras, there were few farmers, horse breeders, or trainers who were without the liniment.[51]

James Snowden, appointed one of the first associate judges in the Northwest Territory in 1799, had no special fondness for horses. He, quite frankly, disliked them and made no secret of the fact. Rather than ride one of the four-legged animals, he walked through the untamed forest in order to hold court and conduct hearings. Once, in a moment of weakness, he was persuaded to ride what was promised to be a calm and smooth-riding horse. Nevertheless, as soon as he was out of sight, he dismounted and, looping the reins over his arm, completed his journey on foot. Deep in thought as he strode along, the judge failed to notice that at some point the horse slipped his bridle. It was not until he reached his destination that he discovered the empty bridle dangling on his arm.[52]

John Eicholtz, classified by neighbors as an eccentric genius, either had no time for his horses or cared very little about them. After his death, according to a newspaper account, his estate sale drew an estimated crowd of five thousand, most of whom were curiosity seekers. Nearly every item was in a state of decay. Some of the ten thousand bushels of corn were twenty-one years old. There were agricultural implements of almost every kind. An ancient wagon had solid wheels cut from a log. A number of old horses were sold that never had seen a halter or a harness. One brought ten cents. For over eight years, another horse had been confined to his stall in a dark barn and was blind as a result.[53]

Pierre Dugan, a Frenchman who lived with his Indian wife and their children on the prairie in present-day Salem township, Champaign County, was a trapper. Joseph Antrim, in his 1872 history of the area, tells a story about Pierre and a horse.

Pierre once purchased a bag of cornmeal from a miller in Kings Creek. Because the Frenchman had no horse of his own, the miller offered him the use of one of his horses. The horse, whose name was Gopher, was a small animal. Pierre thoughtfully looked at the horse, then at the bag of meal, and then at himself. He "took the bag of meal upon his own shoulder and deliberately leading Gopher to a stump, he mounted his bare back, saying as he did so that 'he could carry the bag of meal and the horse could carry him' and in this way he rode home."[54]

Always, there were people who delighted in trading horses. Even at camp meetings, those religious gatherings where people listened intently to impassioned sermons, men found opportunities to indulge in racing and in trading their horses. Between sermons, the men would withdraw a short distance to match the speed of their favorite horses. Many an animal changed hands at these worship services. Some of the pious members in the congregation objected to the hilarity and sternly rebuked the revelers. Wives scolded husbands because they were more interested in "swapping horses than in exposing their souls to salvation."[55]

There was a newspaper account about a camp meeting which was attended not only by a large throng of settlers but also by a group of Wyandot Indians. An observer noted that, including the adults, children and babies, ponies and dogs, the Indians outnumbered the whites by a wide margin. In the midst of a Sunday discourse some of the men were racing their horses not far away. Suddenly, the Indians began whooping, convincing those listening to the preacher that they were being attacked and possibly would be scalped. Reassured such was not the case, they learned that the man who had won the race, and his supporters, merely were expressing their jubilation.[56]

Camp meetings attracted huge crowds. People traveled from far and near, camping in the woods for days at a time while they received spiritual sustenance, renewed old acquaintanceships and made new friends. Mr. and Mrs. H. S. Knapp, who lived in the western part of the state, started soon after daybreak one morning for a camp meeting in Columbus Grove, Ohio. The roads were in a deplorable condition—mud hole after mud hole. The buggy often sank to its axles in spite of Mr. Knapp's attempts to weave his way around the obstacles. Notwithstanding his efforts, the buggy upset and man and wife landed in the mud. Mr. Knapp at length was able to free himself and the horse but could not move Mrs. Knapp. Distraught, he backed-up the horse as close as he could and urging his wife, who had burst into tears, to hold onto its tail, he pulled her out.[57]

A man in Clark County made use of his horse in somewhat the same fashion. James Galloway purchased a one hundred seventy-five pound anvil in Cincinnati and in order to get it home, devised a make-shift sled from the fork of a tree. He loaded the anvil onto the sled and pegged it securely in place. Lacking a harness or rope of any kind, he tied the sled to the horse's tail and, mounting the perplexed animal, rode home.[58]

Story telling, especially stories about one's horses, provided much entertainment in early Ohio. A tall tale that was widely circulated concerned "a Colonel who said that as a young man he had lived on a farm. 'The farm had been well-wooded and the stumps were pretty thick. But we put the corn in among them, and managed to raise a fair crop. The next season I did my share of the plowing. We had a sulky plow and I sat in the seat and managed the horses, four as handsome bays as ever a man drew rein over. One day I found a stump right in my way. I hated to back out, so I just said a word to the team, and, if you will believe it, they just walked that plow right through that stump as though it had been cream cheese.' Not a soul expressed surprise. But

Major S., who had been a listener, remarked quietly: 'It's curious, but I had a similar experience myself once. My mother always made our clothes in those days, as well as the cloth they were made of. The old lady was awful proud of her homespun—said it was the strongest in the state. One day I had just plowed through a white-oak stump in the way you speak of, Colonel, but it was a little too quick for me. It came together before I was out of the way and nipped my trousers. It felt mean, I can tell you, but I put the string on the ponies, and, they just snaked that stump out, roots and all. Something had to give, you know.' "[59]

Horse trading, as one might suspect, took place not only during camp meetings but any place at all where two or more persons chanced to come together and reach an agreement. "One reason why there were so many poor horses around was that people swapped them too often and they did not have time to get fat," was a standard joke of the times.[60]

As early as 1800 horses had become a source of revenue for a few dealers in Ohio. Edmund Martin Jr., at a relatively young age, became a well-regarded judge of horses and at regular intervals shipped horses to the Boston, New York, and New Orleans markets. In 1857 he sold a shipload of horses overseas "to supply the armies of Russia and Turkey then at war." More of his horses found their way to buyers in South America.[61]

The journey southward to New Orleans could be fraught with danger, both going and coming. More than one man drowned on the way down river or was murdered for the money he was carrying and/or the horse he was riding when returning to Ohio.

Lame, old, and broken-down horses frequently were turned loose as immigrants made their way westward. Johnny Appleseed, the eccentric wanderer who planted apple seedlings as he roamed in Ohio, is said to have gathered those horses together in the fall and bargained with a nearby settler for their feed and shelter until the next spring when he led them to good pasture for the summer. If the horses recovered, Johnny would loan or give them away, but "always with the condition of their good usage."[62]

Stealing horses was a serious crime but thieves took the risk and struck anyplace and everywhere. Church meetings were no exception. *The New Era*, a St. Paris, Ohio, newspaper printed in its September 22, 1881, issue: "Neal Mahan lost a horse and buggy at the Carysville camp meeting Monday night."

George Gideon inserted an advertisement in an 1839 paper offering a reward for the return of his horse after it was stolen at a camp meeting.

Stop the Thief.

WAS stolen from the African Camp-meeting, near Mechanicsburg on the night of the 25th inst., a

Dark Bay Mare,

about 7 years old, 16½ hands high, stout built, and well proportioned. When taken she had on a new Spanish Saddle, with brass Horn and Stirrips, Bridle and Martingales. Any person returning said Mare, or giving such information as will lead to her recovery, shall be handsomely rewarded, and all expenses paid.

GEORGE GIDEON.
Woodstock, Champaign Co.
August 27th, 1839   19-4w*

[63]

It was the nearly universal opinion that Ohio was infested with horse thieves and recovering the stolen animals was next to impossible. Some communities became so concerned they formed "Anti-Horse Stealing Societies."

Ordinarily, if horse thieves were caught, the penalties were severe. In 1810, however, two horse thieves incarcerated in a log jail managed to avoid any kind of chastisement other than spending a few days in jail. While awaiting trial, the men entertained the inmates and the jailor and his family with a large repertoire of songs. They sang, not particularly well, but frequently and loudly. One day, following a lengthy performance, it became suspiciously quiet and upon investigation, it was discovered that the two men had made their escape. The musical recital had been a gambit to mask their noisy activity. The thieves never were recaptured.[64]

Corporal punishment, steep fines, and imprisonment were not out-of-the-ordinary. From 1788 until 1803, while Ohio still was governed by territorial laws, it was customary to punish horse thieves with a public whipping. The culprit was obliged to stand with his back bared and his up-stretched arms tied to the whipping post. Sometimes as many as fifty-nine lashes, one hundred lashes in a few localities, were administered for the first offense and from one hundred to two hundred lashes for a second and subsequent offense. A territorial law decreed that if found guilty of a third offense for stealing horses, a prisoner's ears would be cropped. To the best of anyone's knowledge, no such penalty ever was carried out in Ohio on a private citizen.

Individuals found guilty of serious crimes, besides stealing horses, occasionally were sentenced to "Moses' Law," i.e., forty lashes with a whip less one. Offenders of a relatively minor nature might be ordered to carry on their backs "a flag of the United States," i.e., thirteen lashes with a whip.

Whipping, or "flogging" as it sometimes was called, usually was done with "cats." "Cats," an abbreviated term for "cat-o'-nine-tails," consisted of a whip on the end of which was fitted a handful of narrow, rawhide straps. The end of each of the straps was tied in a knot. Depending upon the force with which the whip was applied, the ordeal could be agonizing.

In the spring of 1834, thirty-one years after Ohio was admitted as a state, a man in Van Wert County who stole a horse was arrested, tried, and convicted. Thinking he was to receive a public whipping, he was surprised at the actual punishment. He was stripped naked and tied to a tree. There he remained through the night, exposed to the hordes of mosquitoes.[65]

Whippings were not necessarily administered all at once. A man in Licking County was scheduled to receive "fifty stripes, well laid on (as the law then required), . . . five in the morning, fifteen at noon and thirty the following day at noon." The scene was described in *The Ohio Frontier*. "A circle of about sixty feet diameter was drawn and a cordon established that kept back the crowd that pressed to the line. The prisoner was brought out from the log jail and secured by his upraised hands to the big staple. The first blow of the 'cowhide' simply left a welt. 'A little harder,' said the sheriff," and the four succeeding blows showed in "distinct red lines on the poor fellow's naked back. He received this first installment of his sentence without an audible groan; but when returned to the same position for the second, his utterances and screeches from the first stroke were heart-rending, and when returned to the prison, his audible lamentations and prayers for annihilation before another day were fearful and most painful to be heard. Yet he stood the whole punishment, receiving the following day the heavy remainder of the infliction, and returned to his prison with his back lacerated and bleeding from his shoulders to his hips."[66]

A chapter in Henry Howe's *Historical Collections of Ohio* relating to Lawrence County included a Court of Common Pleas case of "false warranty in the sale of a horse," a case that came to be known as "the case of the bewitched horse." The plaintiff accused the defendant of knowingly selling him an unsound horse. A witness who was present at the time the transaction took place verified that the defendant did, indeed, know the horse was in poor condition. The defendant, however, responded that the horse "was in no way diseased, or in unsound health, but that the drooping appearance arose from his being bewitched, which he did not call unsoundness, and so soon as they could be got out of the horse he would then be as well as ever."[67] Although it would have been of interest, the verdict in the case was not revealed in the account.

Indians were notorious for stealing horses. The Delaware, Shawnee, and other Woodland tribes in Ohio had made use of the horse since the mid-1700s and as trophies of war took horses from their enemies. Bands of roving Indians and warriors on raiding parties stole settlers' horses whenever the opportunity presented itself. Settlers, on the other hand, reciprocated.

Early Ohioans, suspecting there might be Indians in the vicinity, sometimes at night staked-out a horse with a bell around its neck. Concealing themselves in the underbrush, the men would wait, hoping the tinkling of the bell would suggest to Indians intent upon stealing horses that one was nearby. Occasionally, it was successful; oftentimes, the Indians were not deceived.

Simon Kenton was in the act of "reclaiming" some of the Shawnee Indians' horses when he was taken captive. As soon as the Shawnees realized who it was they had caught, one of their most formidable adversaries, they were elated. Kenton's clothing was removed and, according to Edna Kenton, a great niece who wrote about the experience, "they brought up from among their recovered horses a wild, unbroken, three-year-old colt, to which with laughter and difficulty they bound their prisoner. They fastened his hands behind him and his feet under the colt's belly. A halter was passed about his neck and its ends fastened to the colt's neck and rump. Then, all made ready, with Kenton powerless to ward off branches and underbrush from his face and body, they gave the colt a smart blow as they released it and as it dashed off they roared with mirth at the spectacle. The colt pitched, reared, and rolled to rid itself of its burden; the ragged bushes tore its rider's legs and feet; the tree limbs raked and scourged his face and body. If he dropped forward on the colt's neck to avoid being blinded by branches, his back was lashed by them. Every leap of this ride was a hairbreadth escape, for if he once lost his balance he was finished—the halter would hang him. But Kenton knew horses and he bore a charmed life; gradually the animal grew weary and eventually fell in with the party of its own accord and went on quietly enough."[68]

Taking great care that none of their ill-treatment should cause their prize captive's death, he was taken to the Indian village of Chillicothe, the present-day site of Oldtown, Ohio, where he successfully ran the gauntlet. He survived the gauntlet but, nonetheless, was condemned to death and his face blackened in preparation for being burned at the stake. When an old friend, Simon Girty, interceded in Kenton's behalf, the Indians relented and allowed him to be sent to the British Commandant at Detroit where he contrived to escape and make his way home.[69]

Near the end of his life, Kenton resented being referred to as a horse thief and would say, "I never in my life captured horses for my own use, but would hand them over to those who had lost their animals by Indian thefts, nor did I ever make reprisals upon any but hostile tribes, who were at war against the white settlers."[70]

Daniel Boone, the well known contemporary of Simon Kenton, made an escape from his Shawnee Indian captors through a clever ruse with the tribe's herd of horses. Daniel had been adopted by Black Fish, the leading civil chief of the Cha-lah-kaw-tha tribe of the Shawnees and was given a certain amount of freedom. Four months after his acceptance into the tribe, Black Fish sent Boone to locate and drive back to the village the horses that had been turned loose to pasture in the forest. Boone found them by listening for the bell that hung around an old mare's neck. He removed and hid the bell and, returning to Black Fish, reported that he'd been unable to find the horses. The chief again ordered Daniel to go to the forest and this time not to return until he could bring the horses with him. The deception provided Boone with more time in which to travel undetected before the Indians pursued him. In fact, it later was learned, so much time had elapsed the Indians felt certain he had made good his escape and did not attempt to trail him at all.[71]

There were people who later questioned why Boone did not make use of one of the Indians' horses for his escape. He explained that a trail left by a horse might easily have been followed and he could not have backtracked and covered his prints were he not on foot.[72]

Jonathan Pointer was a slave who had been captured by a Wyandot war party and thereafter resided with the family of Tarhe (The Crane), a chief of the Wyandot nation. Near the close of the War of 1812, Jonathan developed the habit of stealing teamsters' horses. While the wagoners encamped for the night, Jonathan would surreptitiously make-off with the horses and conceal them. Openly returning to the camp the next day, he would offer to find their horses—for a price. Tarhe soon learned of the man's duplicity and "told him that if he ever heard of his playing any more such tricks upon travelers, he would remand him back to his master in Virginia. This had the desired effect and Jonathan ceased to speculate in that direction."[73]

John Hamilton, during the War of 1812, was captured and adopted by an Ottawa Indian family. Until his exchange nearly a year later, he endured the same privations and hardships as the Indians. There was a period when the warriors of the village were away preparing for a siege of Fort Meigs and, as a result, food for their families was scarce. Before long, they reached a state of near-starvation. They reluctantly slaughtered a few of their horses, which, John learned, was a normal procedure in times of need. Everyone except Hamilton resorted to eating the horse flesh.[74]

Much like the Indians' and the teamsters' horses, Ohio settlers' livestock usually grazed at-will in the forest and sometimes became lost. Owners offered rewards for their return and, routinely, notices were posted that horses had been found. The following is a typical 1813 example:

"Taken up by William Garvin . . . a dark bay horse, about 14 hands high, supposed to be 12 years old last spring, has a white stripe round her left ham, appears to have been cut with a cord, has some saddle sores, had a pair of old shoes on before, appraised to 22 dollars by Wm. Kenton & Jeremiah Thomas. I do certify the above to be a true copy from my Estray book.
                                                Thomas M. Pendleton, J. P. "[75]

Occasionally, horses approached too close to sinkholes in swampy areas and would become trapped. After having lost several horses in this manner, one family in west central Ohio in the vicinity of Kiser Lake said they finally learned to attach boards to the bottoms of their horses' hooves to keep them from sinking into the morass.[76]

Although the six-horse hitch Ebenezer Folkerth drove would not sink, they would stall in the marshy fields on his Montgomery County farm pulling a light wagon loaded only with a barrel of flour.[77]

During the War of 1812, Colonel Richard Mentor Johnson, the man who was credited with killing the famous Indian, Tecumseh, marched his mounted troops north through the Black Swamp toward Detroit to do battle with the British and their Indian allies. He reported that the expedition often halted as equipment and horses floundered. Some horses were left to die. Colonel Johnson himself was forced to leave behind two riding horses that became hopelessly mired.[78]

In parts of the state where one found patches of ground called "barrens," there occasionally were unexpectedly treacherous sinkholes where horses bogged down chest-deep. If found in time they could be pulled out; otherwise, they starved to death or were killed by wild animals.

Accidents with horses were not uncommon. A March, 1880, newspaper told about a team of horses that drowned in Mad River. Richard Bean was driving a loaded wagon along the bank of the river when the horses began backing-up. They backed too far and went over the embankment. The driver jumped to safety and was not injured but the horses, pinned by the wagon and hampered by the harness, drowned.[79]

H. R. Cline of Casstown, Ohio, drove to the woods to cut a tree and, tying the team a safe distance away, commenced chopping. He completed cutting the tree but, instead of falling where he had anticipated, it fell the wrong way, landing on the horses and killing them both.[80]

An 1888, weekly newspaper reported, "John Igou lost a valuable driving horse. It was kicked by a strange horse, he had just purchased, on the ankle. The injury was so great that the animal had to be killed."[81]

Lightning frequently killed horses while they stood in a field or under a tree. Ora Hess' horse was killed inside his barn. A blinding flash of lightning struck the barn and followed a metal water pipe to the ground where it killed the horse as it drank from a trough. The Hess family was especially hard hit for the same bolt instantly killed their eight-year-old son. Mr. Hess, who was walking on the far side of the boy, received a tremendous jolt and his life hung in the balance for many months.[82]

The *New-Era,* a St. Paris, Ohio newspaper, printed the following account on November 3, 1881.

"RESURRECTED ON EARTH

Our basket maker, supposed Wednesday that his horse had died. After carefully digging a grave, and calling his wife to help deposit the remains of the quadruped, they both took hold to consign the dead animal to its last resting place when behold, the dead brute lived, moved and breathed, and the old woman who had been assigned to duty at its latter extremities, was knocked out of time, and the horse jumped up and made for the stable, and is now living in good health, almost a victim of suspended animation."

Another paper carried the item: "A horse that was in the habit of scratching his nose with his hoof got its hoof in its mouth and it remained there one whole night before the animal was relieved from its uncomfortable predicament."[83]

People were killed in horse-related accidents. James Swisher was kicked by a horse and died in 1855.[84] Mrs. James Noble, of Fayette County, mother of eleven children, died as a result of a runaway horse in 1864.[85] A little boy was killed when a horse-drawn cart ran over him. The notice read: "killed by a cart wheel passing over the head." A second report was considerably more explicit: "The wagon wheel passed over his head, crushing it to jelly."[86]

On an 1830 tombstone in a cemetery in Licking County is the engraving: "Oh, think not that you are safe when in your health. The kick of a horse was the means of my death."[87]

A death notice for Nelson Rhodes explained precisely what had happened when the horse he was driving took fright. "Esquire Nelson Rhodes and Colonel S. H. Hedges started on a buggy ride. Rhodes was driving a spirited young horse he recently had purchased. They drove east from North Main on Church Street to East Lawn and then to Boyce. The horse became frightened of some children playing nearby and ran to the side of the street, striking a fence and tearing off a board which became entangled in the gears of the buggy. One end protruding and just reaching the horse's side where every motion the horse made the board strike the horse, frightening him more. He crossed to the other side of the street and ran up a bank. The buggy turned over, throwing both men out of the buggy. Colonel Hedges struck a fence with such force to break a board. Rhodes was thrown against a tree and his skull smashed."[88]

Not everyone was privileged to own docile, well-behaved horses. Many horses had less desirable traits. Some were too nervous and excitable and shied at sudden noises or movements; some were vicious; some had an almost inexplicable tendency to run away with their riders and/or drivers; and some were just plain balky. These were the animals most likely to cause mishaps. Early newspapers were full of such accounts.

In 1861, there was a young man in Stark County who, while on his way to enlist in Co. A of the 8th Ohio Infantry, met with an accident and broke his leg. His skittish horse became frightened and threw him and the buggy he was driving over an embankment. Prevented from enlisting, he enrolled in medical school. Upon completion of his training a year later, he succeeded in enlisting in the Union army and served until the end of the war.[89]

An 1879 newspaper printed an item about a horse who fell on a man: "Charles Wood met with a severe accident last week by a horse falling on him, but he is recovering."[90] Another issue of the same paper told: "Two horses hitched to a wagon, during a temporary aberration ran up Scioto Street with more force than safety, last Monday. A smashed wheel, a bruised driver and a few caroms on the other wagons and the tale is told."[91]

A four-horse hitch, frightened at something on the sidewalk in a small Ohio town, ran away. "It made respectable kindling wood of the wagon." wrote the attendant reporter[92] A horse pulling a sleigh one snowy night was startled by a shrill train whistle. It took off on a dead run, upsetting the sleigh and seriously injuring one of the occupants. "The sleigh was badly demoralized" read the notice in the paper.[93]

One Sunday morning in 1872, a family was thrown from their wagon when the two-horse hitch took fright and dashed up the street. Bystanders ran in pursuit, finally ending the chase in the middle of town where the team crashed into a hat store window. "The damage to the store, the wagon, and the horses was considerable."[94]

A team of horses, stung by bumblebees when they stepped on the nest, reared and, out-of-control, ran wildly across the field, stopping only when they plunged into a fence. Badly entangled in the harness, the fence, and the harrow, they stood quivering, fitfully stomping their feet. The boy who had been driving the team ventured to calm the frightened animals and was attempting to remove their harness when one of the horses took a sudden step backwards. The boy couldn't move out of the way fast enough and the horse stepped on him, breaking one of his legs.[95]

Accidents were taken for granted. In 1797, three families from Vermont left their homes for what today is Montgomery County, Ohio. They journeyed together in two, large, three-horse wagons. Moving along at a brisk trot one day, there was a sudden jarring bump. A man fell from a wagon and was thrown under its wheels. Before the team could be stopped, both wheels ran over him. The only comment in a Montgomery County biography regarding the accident was: "Such bold spirits were not to be thwarted by a little accident like that."[96]

Not quite everything that befell horses could be attributed to accidents. To illustrate—pioneer weddings were great social events and to be excluded from the guest list was viewed as an insult. Uninvited individuals were known to take offense and for revenge, cut off the manes, forelocks, and tails of the horses belonging to the people in attendance.

Young Joseph Barker, one spring day, was whitewashing the door of his house when an old red mare ambled up to the front gate and stood quietly eating grass. It had become an almost daily occurrence and one the family did not appreciate. She was an intractable animal who would not willingly move away. Her owner, a next door neighbor, was an ill-natured soul who refused to keep the mare corralled.

On this particular morning, Joe decided to play a joke and gave the horse a coat of whitewash. When the mare eventually wandered home, her owner did not recognize her and drove her away. The horse returned a second time to her owner's yard. Still not recognizing her, the man called to his dog to chase her away and followed along cursing and pelting her with rocks. Although the villagers were aware what had taken place, they professed to have no knowledge of the horse when, after a few days, the horse's owner decided that she had been stolen. Only when patches of the whitewash had rubbed off, revealing her true color, did the owner realize the mare was his.[97]

In Ashland County in 1812, a rash act on the part of a white man had dire results and escalated the war between the Indians and the encroaching whites. Horses that belonged to a small party of Indians, turned loose to graze, destroyed the crops of Frederick Zimmer. In fury, the man tied "clap boards" to the horses' tails, i.e., he fastened boards to their tails in such a manner that as the horses ran, the flat surface of the boards snapped together with a sharp, explosive-like sound. Zimmer finished attaching the devices, slapped the horses on their rumps, and watched the terrified animals run through the woods. (One source says the boards were set afire). The Indians were not amused. Their wrath was so great they massacred and scalped the Zimmer family and a neighbor who was hurrying to warn them.[98]

Fifty years after the war with the Indians in Ohio came to an end, three Ohio generals rode horses that became famous during the Civil War. All three generals, who fought on the side of the Union army, were skilled horsemen from a young age.

Ulysses Simpson Grant (1822-1885), a short, small boy, had to climb into a manger to put a bridle on a horse and stand on a box to throw a harness over a horse's back. Once, in his father's absence, he led an unbroken colt from its stall and, hooking it to a sled, hauled firewood from the woods using only a single line. The colt performed as if he'd always been accustomed to the task.

During much of the fighting in the war, Grant rode half a dozen mounts but the most notable was Cincinnati, so named because the patriot who gave the horse to Grant was from Cincinnati. The spirited dark bay was a well-bred animal that stood seventeen hands high. He was a favorite of General Grant who permitted only two other people ever to ride him—a friend who'd saved Grant from drowning as a boy and President

Abraham Lincoln who rode Cincinnati when he reviewed the troops at City Point during the final month of the war.[99]

General William Tecumseh Sherman (1820-1891), who was born in Lancaster, Ohio, had two horses of which he was especially fond, Sam and Lexington. Sam was a big, powerful bay who stood quietly under fire, so quietly that Sherman could write orders while sitting astride the horse. Lexington was a Kentucky thoroughbred and the horse the General rode in Washington D. C. at the last review of his army following the close of the Civil War.[100]

General Philip Henry Sheridan (1831-1888), reared in Somerset, Ohio, was a fiery little man who, during the Civil War, became known as "Little Phil." He rode a black Morgan horse called Rienzi. Mounted on Rienze, the General raced along the road from Winchester, Virginia, when fighting broke out at Cedar Creek, rallying his men who were retreating. Urging them to turn and fight the oncoming Confederates, the men took heart, fought back, and ultimately won the fight. This remarkable, eleven mile ride resulted in the horse's being renamed "Winchester." The ride was immortalized in Thomas Buchanan Read's poem, "Sheridan's Ride."[101]

Some horse stories are of interest only because of the people with whom the animals were associated. A case in point was Confederate, John Hunt Morgan (1825-1864) and his band of volunteer cavalrymen who earned the appellation "Morgan's Raiders" during the Civil War. From July of 1862 when Morgan broke through federal lines in Kentucky, until June of 1864 when he was killed, he and his men swept through northern Kentucky, southern Indiana and east through southern Ohio raiding, burning bridges, capturing railroad supplies and, everywhere they went, "liberating" horses. As their horses gave-out, they simply commandeered others.

While dashing across Ohio in 1863, the home of a family by the name of Davis lay in Morgan's path. The Davis' ten year old son was plowing corn when three of the rebels approached and demanded his horse. "Young Davis, not wishing to be interrupted, was about to proceed when his 'Get up, Joe!' was responded to by the rebels leveling their revolvers upon him in a decidedly suggestive manner. Changing his mind with a 'Whoa, Joe!,' he permitted them to take the horse, and this, along with two others, were never recovered."[102]

The early Ohio saga of Ann Sargent Trotter Bailey, better known in the Gallipolis, Ohio, community as "Mad Ann," began in 1700 in Liverpool, England, where she was born and named in honor of the queen. One version of her story says she was kidnapped at age fourteen and brought to the United State where she was sold to defray her expenses. A different version says she was married at age thirty to John Trotter, a soldier, and came here with him. John was killed by Indians at the battle at Point Pleasant in 1774. Ann swore vengeance on all Indians for his death and actively aided his former comrades-in-arms by spying and carrying messages on her black horse, Liverpool. Liverpool, named for the town where she was born in England, carried her safely through many harrowing situations.

Ann once rode one hundred miles on Liverpool to replenish a garrison's ammunition supply. On another occasion when she was being pursued by Indians, she rode into a thicket where she was forced to dismount and leave the horse for the Indians to capture. She crept into a hollow sycamore log where she remained concealed as the Indians searched for her. After they had gone, she followed their trail and in the darkness of night, retrieved Liverpool. Mounting him when they were safely out of gun range, she gleefully shouted her defiance and raced away. Eventually, the Indians believed her insane and bothered her no more.[103]

Volumes have been written about Tecumseh, the charismatic Shawnee Indian who worked to unite all Indians into a single alliance that would defend their lands against the white man's intrusion. Tecumseh's friend, Jonathan Alder, delighted in telling a seldom heard tale about Tecumseh and a horse trade.

"I was well acquainted with him [Tecumseh]. I sold him a keg of rum one day for a horse; the horse got sick and died, and shortly afterward I told him he ought to give me another horse. He said he had drunk the rum up and it was all gone, and he supposed I was about as well off as he was. He said the rum was of no use to either of us, and that he had suffered all the bad consequences of drinking it. He reasoned that the horse had done me as much good as the rum had done for him, and perhaps more, but as it was, if I was satisfied we would quit square, and so we did."[104]

After the War of 1812 relatively few Indians remained in Ohio. The last of the surviving Indian tribes to live together in organized groups in the state, the Wyandots, were forced to resettle in the far west in the spring of 1843. Colonel John Johnston, a respected Indian agent at Piqua, Ohio, helped with the negotiations and although the agreement was concluded in March of 1842, preparations required a full year. Over seven hundred ponies had be shod. Three blacksmiths in Rushylvania, who were among those employed to accomplish the task, told what a formidable assignment it became. In order to shoe them, each one of the animals physically had to be thrown to the ground and tied down.[105]

To assist with the move, wagons and teams of horses were hired from settlers in the area. In the end, there was a total of one hundred and twenty wagons and buggies. These were divided into companies of from thirty to forty with an overseer or chief in charge of each unit. Nearly three hundred people rode horses or walked along behind. One account said seven hundred fifty Wyandots made the trip. People lined the streets in every town through which the Wyandots passed to watch their poignant and unforgettable departure.

As villages and towns grew and trade developed, the need for communication between the different communities arose. Primitive trails became roads. Horses carrying riders, horses pulling wagons and horses hitched to buggies and carriages were seen with more and more regularity. Hitching posts and hitching rails were placed around the perimeter of town squares and in front of residences and stores. Horses were in evidence everywhere. They had become the prevailing form of transportation.

Sooner or later, it was inevitable that local governing bodies would enact laws relating to horses. One small, west central Ohio town, passed an ordinance in 1858 that read in part: "It shall be unlawful for any person or persons, to leave any horse, mare, gelding, mule, or ass, standing in any street, alley, or public ground, in the corporation, unless the same be securely hitched or fastened, or some suitable person be left in charge thereof; and no person shall hitch or fasten any horse, mare, gelding, mule or ass, to any ornamental or shade tree, or tree box, in any street, alley, or public ground within the corporate limits, or to any public lamp post, or to any private lamp post, without the consent of the owner thereof; and any person who shall in any way violate any, or either of the provisions of this ordinance, shall upon conviction thereof, before the Mayor, be fined in any sum not less than one, nor more than ten dollars, for each and every offense, at the discretion of the Mayor."[106]

Villages often assessed fines for riding horses too fast or driving buggies too rapidly across bridges.

Cemeteries found it necessary to post notices like the following: "Walk your horse." Visitors on horseback were cautioned to confine their walks "to the avenues and paths."[107] Signs that read, "Refrain from tethering your mounts to trees and shrubs," frequently were seen.

The phrase, "at the end of your rope," is thought to have originated from the practice of tethering, or tying, one's horse. While tied, an animal could graze, moving ever farther and farther away until, eventually, the rope was extended to its full length and the horse could stretch no farther.

As Ohio's population increased, so, too, did the number of horses. In 1800, three years before Ohio achieved statehood, the area that comprised what was to become Ohio had an estimated population of 45,000. By 1810, it had grown to 230,760 and was ranked as the thirteenth most populous state in the nation. In 1860, the population leaped to 2,339,511 and, according to the *Nineteenth Annual Report of the Ohio State Board of Agriculture,* there were 675,644 horses in 1865.[108] By 1891, there were 3,672,329 inhabitants and 846,789 horses.[109]

By the mid 20th century the population totaled over 5,759,394[110] but the number of horses had diminished to 504,000.[111] Following the end of World War 1, due primarily to the expanding popularity of automobiles, interest in horses waned. The only horse that remained in favor was the saddle horse.

Today, the 21st century, more people are riding horses than ever before. Horse enthusiasts have taken to heart a statement made by the respected Ohio State University Professor and noted horse judge, D. J. Kays: "There isn't anything so good for the inside of a man as the outside of a horse."[112]

****

The use of mules by early Ohioans must not be underestimated. Many pioneers preferred them. Mules, the result of the unique cross between a mare and a jack, were faster than oxen by a factor of three to five miles a day but slower than horses. They cost nearly twice as much as a yoke of oxen but were less expensive than horses. They were sturdy, heavy-muscled, sure-footed animals, well suited for hard labor and considered indispensable for road work, lumber camps, railroad crews, and for different types of army service. Although they matured slowly, they lived to a great age.

Mules also were notoriously stubborn and contrary. Some of the species were downright truculent. The June 17, 1880, *New Era,* a St. Paris, Ohio, weekly newspaper, printed the following: "Francis Heath, a young man of Concord township, had his nose partly torn off and an eye so badly injured that he probably will lose it, by being kicked by a vicious mule, from which he was trying to make his escape."[113]

George Washington was responsible in large measure for encouraging the production of mules in the United States. When the King of Spain sent him a jack and two jennys in 1787 and his close friend, the Marquis De Lafayette, also sent him a jack and two jennys, Washington began promoting them in earnest.[114] Slowly but surely the strain gained in popularity. By 1842, in addition to nearly constant demand in this country, there was a steady trade in mules with the East and West Indies. Ohio alone was estimated to have 22,228 mules in 1867.[115]

****

A human interest story in a history of Belmont County revolved around coal miner, Jake Heatherington, and his three and one half foot tall donkey, Jack. In 1837, Jake purchased a piece of land on credit and began mining coal. Jack was his only helper. Jake bought more land, lots more land, and mined more coal. Many of the steamers that made their way up and down the Ohio River were fueled with Jake Heatherington's coal.

Jake prospered. He bought a glass-works factory, more than thirty properties, and his own steamboat. In 1870, he built an imposing residence overlooking the river. A keystone, a replica of Jack's head, graced the arch above the front entrance and the house became known as "The House That Jack Built." On the day the house was completed, Jake led Jack up the steps and through the house, room by room, talking non-stop how the two of them had worked side-by-side and that all their hard labor had culminated in this fine house. Several years later, at the age of forty, the little donkey died. Jake buried him under an apple tree close to the house.[116]

****

Early in Ohio's history, enterprising land speculators with high expectations of selling the newly-opened lands, acquired immense tracts. As a way of enticing people to purchase the land, agents for the various companies distributed handbills throughout the East, primarily in Virginia and the Carolinas, magnifying the wonders of the new country. Their claims frequently were preposterous. Drawings illustrated flax bearing pieces of fabric in place of leaves; springs flowing with brandy instead of water; maple trees dripping sugar; swamp plants that yielded candles; and custard that sprouted from a shrub. The swamp plants in reality were cattails and the custard referred to the soft, banana-like pulp of pawpaws.[117]

There was published, however, one negative interpretation of life in Ohio. The leaflet depicted a plump, well-dressed gentleman on a sleek, fat horse with the legend "I'm going to Ohio," meeting a man going in the opposite direction who was dressed in rags and riding a mere skeleton of a horse. Underneath was printed, "I've been to Ohio."[118]

Chapter 8: OHIOANS AND THEIR HORSES

1. Henry Howe, *Historical Collections of Ohio,* vol. 2 (Cincinnati, Ohio: Krehbiel & Co.,1904),p 274.
2. *History of Preble County* (Cleveland, Ohio: H. Z. Williams & Bros., 1881), p 175.
3. William Steele, *Old Wilderness Road* (Orlando, FL: Harcourt, 1968), p 153.
4. R. E. Dills, *History of Greene County* (Evansville, IN: Unigraphic, 1881), p 748.
5. Deborah Halverson, "Journey of Jonathan Hale, *Early American Life* (June, 1980): pp 20-76.
6. Ray Crain, *Land Beyond the Mountains* (Urbana, Ohio: Main Graphics, 1994), pp 73-74.
7. *Ohio Historical Review* (Columbus, Ohio: pub./ed., Ralph S. Estes, n. d.).
8. Henry Howe, *Historical Collections of Ohio,* vol. 2 (Cincinnati, Ohio: Krehbiel & Co.,1904), p 723.
9. W. H. Beers, *History of Madison County* (Chicago: W. H. Beers & Co., 1883), p 284.
10. R. S. Dills, *History of Fayette County* (Evansville, IN: Unigraphic, 1881), p 945.
11. Henry Howe, *Historical Collections of Ohio* (n. d.).
12. John Christian Norman family records: author's ancestors.
13. W. H. Beers, *History of Montgomery County* (Chicago: W. H. Beers & Co., 1882), p 369.
14. ——, *History of Clark County* (Chicago: W. H. Beers & Co., 1881), p 44.
15. R. S. Dills, *History of Fayette County* (Evansville, IN: Unigraphic, 1881), p 966.
16. Nelson Evans, *History of Adams County* (West Union, Ohio: Emmons Stivers, 1900), p 560.
17. Henry Howe, *Historical Collections of Ohio* (Cincinnati, Ohio: Derby, Bradley & Co., 1848), p 400.
18. W. H. Beers, *History of Champaign County* (Chicago :W. H. Beers & Co., 1881), p 539.
19. ——, *History of Madison County* (Chicago: W. H. Beers & Co., 1883), p. 617.
20. William A. Galloway, *Old Chillicothe* (Xenia, Ohio: Buckeye Press, 1934), p 244.
21. *Ohio Historical Review*, vol. 3, no. 18 (Columbus, Ohio: pub./ed., Ralph S. Estes), p 8.

22. R. S. Dills, *History of Fayette County* (Evansville, IN: Unigraphic, 1881), pp 741-742.
23. A. B. Sears, *Thomas Worthington* (Columbus, Ohio: The Ohio State University Press, 1958), pp 15-16.
24. ——, pp 19-20.
25. ——, p 212.
26. Virginia L. Hunt, "What Went on Around Here," Hunt family manuscript, 1958.
27. *Echoes,* vol.39, no. 5 (Columbus, Ohio: Ohio Historical Society, Oct./Nov., 2000), p 3.
28. Elizabeth Hughes Ward diary, 1814-1838: Champaign County Historical Museum, Urbana, Ohio.
29. Benjamin F. Prince, *Standard History of Springfield and Clark County* (Chicago: American Historical Society, 1922), p 215.
30. R. S. Dills, *History of Greene County* (Evansville, IN: Unigraphic, 1881), pp 274-275.
31. W. H. Beers, *History of Montgomery County* (Chicago: W. H. Beers & Co., 1882), p 164.
32. ——, p 6.
33. William Rockel, *Twentieth Century History of Springfield and Clark County* (Chicago: Biographical Pub. Co.,1908), p 564.
34. W. H. Beers, *History of Madison County* (Chicago: W. H., Beers & Co., 1882), p 524.
35. Francis Parkman, *Oregon Trail* (Garden City, N.Y.: Doubleday & Co., 1849), p 240.
36. Orton G. Rust, *History of West Central Ohio* (Indianapolis, IN: Historical Pub. Co., 1934), p 457.
37. Henry Howe, *Historical Collection of Ohio,* vol. 1 (Norwalk, Ohio: Laning Printing Co., 1896), p 249.
38. W. H. Beers, *History of Clark County* (Chicago: W. H. Beers & Co., 1881), p 743.
39. Ben Riker, *Pony Wagon Town* (Indianapolis, IN: Bobbs-Merrill Co., 1948), pp 211-212.
40. *Champaign Democrat, (Urbana, Ohio),* April 1, 1880.
41. M. E. Ensminger, *Horse Husbandry* (Danville, IL: Interstate Printers & Pub., 1951), p 28.
42. *History of Logan County and Ohio* (Chicago: O. L. Baskin, 1880), p 179.
43. William Dean Howell, *Stories of Ohio* (Cincinnati, Ohio: American Book Co., 1897), p 271.
44. *(St. Paris, Ohio) New Era,* June 13, 1878.

45. David Miller, *Practical Horse Farrier* (Hamilton, Ohio: E. Shaeffer, 1830), pp 120-121.
46. Joseph Ware, *History of Mechanicsburg, Ohio* (Columbus, Ohio: F.J. Heer, 1917), p 27.
47. Henry Howe, *Historical Collections of Ohio,* vol. 1 (Cincinnati, Ohio: Laning Printing Co.), p xviii.
48. Fairfax Downey, *Famous Horses of the Civil War* (New York, N.Y.: Thomas Nelson & Sons, 1960), p 67.
49. *Western Citizen and Urbana (Ohio) Gazette,* September 29, 1840, Urbana, Ohio.
50. *Urbana (Ohio) Citizen and Gazette,* May 24, 1888, Urbana, Ohio.
51. James Young, "For Man and Beast," *Timeline* (Columbus, Ohio: Ohio Historical Society, Sept./Oct.,2001, pp 44-54.
52. R. S. Dills, *History of Greene County* (Evansville, IN: Unigraphic, 1881), p 632.
53. Judge Evan P Middleton, *History of Champaign County, Ohio* (Indianapolis, IN: Bowan & Co., 1917), pp 1139-1140.
54. Joshua Antrim, *History of Champaign and Logan Counties, From Their First Settlement* (Bellefontaine, Ohio: Press Printing Co., 1872), p 427.
55. R. E. Banta, *The Ohio* (New York, N.Y.: Rinehart & Co., 1949), p 348.
56. *Western Citizen and Urbana (Ohio) Gazette,* August 8, 1861.
57. Thaddeus Gilliland, *History of Van Wert County* (Chicago: Richland & Arnold, 1906), p 65.
58. W. H. Beers, *History of Clark County* (Chicago: W. H. Beers & Co., 1881), p 723.
59. *Daily News (Urbana, Ohio),* May 27, 1882, Ohio.
60. *Urbana (Ohio) Citizen and Gazette,* October 10, 1878.
61. W. H. Beers, *History of Brown County* (Chicago: W. H. Beers & Co., 1883), p 170.
62. ———., *History of Champaign County* (Chicago: W. H. Beers & Co., 1881), pp 297-298.
63. *Western Citizen and Urbana (Ohio) Gazette,* September 10, 1839.
64. Edna Kenton, *Simon Kenton, His Life and Period, 1755-1836* (New York, N.Y.: Random House from the Doubleday edition, 1928), p 280.
65. Thaddeus Gilliland, *History of Van Wert County* (Chicago: Richland & Arnold, 1906), p 65.
66. *The Ohio Frontier: An Anthology of Early Writings,* ed Emily Foster (Lexington, KY.: University of Kentucky Press, 1996), p 131.
67. Henry Howe, *Historical Collections of Ohio,* vol. 1 (Cincinnati, Ohio: Krehbiel & Co., 1904), p 57.

68. Edna Kenton, *Simon Kenton, His Life and Period* (New York, N.Y.: Random House from the Doubleday edition, 1930), pp 110-112.
69. ——.
70. ——, p 99.
71. William A. Galloway, *Old Chillicothe* (Xenia, Ohio: Buckeye Press, 1934), pp 258-260.
72. ——, p 260.
73. *History of Wyandot County* (Chicago: Leggett, Conway & Co., 1884), 272-273.
74. Joshua Antrim, *History of Champaign and Logan Counties From Their First Settlement* (Bellefontaine, Ohio: Press Printing Co., 1872), pp 66-68.
75. *Farmer's Watch-Tower (Urbana, Ohio)*, January 13, 1813.
76. Ludene E. Fansler, "Laurence A. Fansler" *Champaign County, Ohio, 1991* (Dallas, Texas, Taylor Pub. Co.), p. 136.
77. W. H. Beers, *History of Montgomery County* (Chicago: W. H. Beers & Co., 1882), p 439.
78. Ray Crain, *Long Green Valley* (Urbana, Ohio: Main Graphics & Associates, 1996), p 68.
79. *St. Paris (Ohio) Examiner*, March 25, 1880.
80. ——.
81. *Urbana (Ohio) Citizen and Gazette*, May 24, 1888.
82. Old newspaper clipping, no name or date.
83. *Urbana (Ohio) Citizen and Gazette*, May 10, 1888.
84. Urbana, Ohio, newspaper, possibly the *Urbana Free Press,* 1855.
85. R. S. Dills, *History of Fayette County* (Evansville, IN: Unigraphic, 1881), p 951.
86. (St. Paris, Ohio) *New Era*, August 21, 1879.
87. Henry Howe, *Historical Collections of Ohio,* vol. 2 (Cincinnati, Ohio: Krehbiel & Co., 1904), p 84.
88. *(Urbana, Ohio) Citizen and Gazette*, September 26, 1878.
89. *Portrait and Biographical Record of Stark County* (Chicago: Chapman Bros, 1892), p 171.
90. *(Urbana, Ohio) Citizen and Gazette*, February 20, 1879.
91. ——, Feb. 27.
92. ——, Feb. 13.
93. *(St. Paris, Ohio) New Era*, Jan. 9, 1879.
94. *(Urbana, Ohio) Citizen and Gazette*, March 21, 1872.
95. *(St. Paris, Ohio) New Era*, August 15, 1878.
96. W. H. Beers, *History of Montgomery County* (Chicago: W. H. Beers & Co., 1882), p 6

97. S. P. Hildreth., *Biographical and Historical Memoirs of Early Pioneer Settlers of Ohio With Narratives of Incidents and Occurrences in 1775* (Cincinnati, Ohio: Derby & Co.,1852), p 435.

98. Robert Price, *Johnny Appleseed: Man and Myth* (Urbana, Ohio: Urbana University, 1956) p 98.

99. William Dean Howells, *Stories of Ohio* (Cincinnati, Ohio: American Book Co., 1897), pp 242-243.

100. Fairfax Downey, *Famous Horses of the Civil War* (New York, N.Y.: Thomas Nelson & Sons, Ltd., 1960), pp 71-72.

101. ——, pp 68-71.

102. *History of Wyandot County* (Chicago: Leggett, Conway & Co., 1884), p 584.

103. Henry Howe, *Historical Collections of Ohio,* vol. 1 (Norwalk, Ohio: Laning Printing Co., 1896), pp 677-679.

104. *History of Madison County*, 1883, pp 286-278, W. H. Beers & Co., Chicago

105. *Memoirs of the Miami Valley,* Vol. 1, 1919, p 305, Leggett, Conway and Co., Chicago.

106. Urbana, Ohio, ordinance, June 11, 1858.

107. Richard Meyer *Cemeteries and Grave markers* (Logan Utah: Utah State University Press, 1992), p 318.

108. *Nineteenth Annual Report of the Ohio State Board of Agriculture, 1864* (Columbus, Ohio: R. Nevins Printer,1865), p xii.

109. *Annual Report of the Secretary of State (Ohio) For 1891* (Columbus, Ohio: The Westbote Co., State Printers), pp 565-567.

110. *Ohio Fourteenth Federal Census 1820-1900.*

111. Ohio Agriculture Statistics Services, Reynoldsburg, Ohio

112. Kays, D. J., *Horse, The,* 1953, p 6, Rinehart & Co., N.Y.

113. *(St. Paris, Ohio) New-Era*, June 17, 1880.

114. H. M. Briggs, *Modern Breeds of Livestock,* 1958, pp 730-736, McMillan Co., N.Y.

115. *Report of the Commissioner of Agriculture, 1867* (Washington D.C: Office of Printers to House of Representatives).

116. Henry Howe, *Historical Collections of Ohio,* vol. 1 (Norwalk, Ohio: Laning Printing Co., 1896), p 322-325.

117. ——, p 669.

118. *The Midwest, A Collection From Harper's Magazine* (New York, N.Y.: Gallery Books, 1991), p 182.

PIGS AND MORE PIGS

Chapter 9

Printed handbills distributed in the late 1700s and early 1800s, designed to attract immigrants to Ohio, had cartoons of roasted pigs running at-large with knives and forks stuck in their backs, squealing, "Come and eat."[1] While the rhetoric was not expected to be taken seriously, it was intended to convey the idea that Ohio was the land of opportunity.

Such tactics may have had some impact for wave after wave of eager people came. Many brought their livestock along with them over the rugged trails. There were cows and sheep, turkeys, ducks and chickens, dogs and cats, and, of course, there were pigs.

Actually, when the first pioneers arrived, wild hogs already were well established in the land. The commonly held belief is that some of the hogs, introduced to the North American continent by 15th and 16th century colonists, escaped from their enclosures, increased in numbers, and in the course of time ranged far beyond any known settlements. Left on their own, the escapees and their subsequent offspring grew wild and in the process developed decidedly unfriendly dispositions.

Wild hogs were considered one of the most savage and ferocious creatures in the forest. With their sharp tusks, which often were several inches long, they defended themselves quite well against predators such as wolves, panthers, and bears. An eight hundred pound boar could bite through an animal's, or a man's, leg as easily as slicing through butter. They were masters at foraging, indiscriminately eating almost anything. They had a predilection for devouring snakes, poisonous and non-poisonous alike. Their attack was sudden and headlong. There were people who testified that these indurate animals could outrun the fleetest horses for a short distance. Their hearing was acute. If a wild hog "heard, a hundred yards away, an acorn rattling, as it dropped through the leaves, he could run and catch it on the first bounce every time," observers claimed.[2]

The hogs were slab-sided, roach-backed, long-bodied, long-legged, had long snouts and were tall. Frequently called "razorbacks" because of their appearance, these fiercely independent and aggressive animals sometimes were known as "elm-peelers," "rail-splitters," "land sharks," "alligators," and "prairie rooters."[3] Pig skin was so tough that it regularly was tanned and used for horse collars and serviceable shoes.

When aroused, the bristles on their backs stood up like the quills on porcupines. One description compared the hogs' lean, flat sides to the elongated, tapering bodies of fish. Even the bristles across the top of the critters' backs somewhat resembled the fins on fish.

About the time Logan County was established in 1817, hogs were becoming increasingly more numerous. Primarily for this reason, one of the first settlers, a keen-witted fellow, suggested the new county should be called "Bristle County."[4]

Hogs often roamed together in herds and when their young were threatened, would close in a protective circle around them. If a wolf or bear ventured too near, they would tear him to pieces with their tusks.

People found it prudent to be on their guard. An early pioneer in Medina County, in the northeastern part of Ohio, told how he had escaped from rummaging wild hogs. Every week or two he, or someone else from his small community, walked to Cleveland to collect the mail and purchase necessary supplies.

With a rifle over his shoulder and a haversack on his back he started out one day. In mid-afternoon, he became aware that he was being surrounded by wild hogs and "discovering a large fallen tree ahead which had turned up by the roots, he hastened to and climbed upon the same, perching upon the high roots some fifteen feet above the ground. He was not a moment too soon, for the hogs had closed around him and some of the old boars, with their tusks protruding from their villainous jaws and the froth dripping from their mouths, attempted to climb up the roots upon which he was perched. He lost no time upon firing upon them whenever he could fire his rifle, which he had to snap eight or ten times for each discharge, because of the preponderance of seeds among his percussion pills. However, he killed a dangerous boar at each discharge. As each one fell, with a slight squeal of distress, the others would go and smell the blood, actually placing their ugly snouts to the bullet hole. They at once began to utter a peculiarly ominous grunt and one by one withdrew from the scene." It was late at night when he reached Cleveland. "Early next morning, he had the lock of his rifle altered, provided himself with proper ammunition, and with his mail and other supplies, started on his return trip, hoping to have a little more experience with the wild hogs." When he reached the place where he had taken refuge from the wild hogs the previous day, he counted sixteen dead boars that he had killed. No live hogs were to be seen.[5]

A family of German emigrants from Pennsylvania resided temporarily in a cabin near today's Olmstead Falls, Ohio, in Lorain County. Shortly after taking up residence, a party of Indians stopped. One of the women gingerly placed her baby in its cradleboard against a tree and joined the others indoors. No sooner had she walked inside, however, than a wild pig dashed from the woods and, grasping the infant in its mouth, turned back into the trees. In a body, everyone ran from the cabin. Arming themselves with clubs, they followed in pursuit and caught-up with the hog who, still clutching the infant in its cradleboard, turned and charged at the group.

Leaping aside, they began striking the infuriated animal. Not until it had been severely beaten did it relinquish its hold on the baby.[6]

Settlers who drove livestock from their former homes to the Ohio frontier generally stopped for the night when they saw the animals were beginning to tire. In order to protect them from wild animals, people slept with their livestock near at hand. Hogs, in particular, often were fed grain each evening in order to condition them to follow the wagons during the daytime. Periodically, however, a few would be lost or fall behind, usually to come trotting into camp several days later. Hogs had remarkable memories and there were some who did not take kindly to leaving their old, familiar habitats.

In the spring of 1807, Robert McFarland's hog returned to its previous home in Kentucky. Enroute to central Ohio, the family made camp for the night not many miles from their final destination. When they awoke in the morning, they found their sow and her seven pigs gone. Two weeks later, they received word that all eight were safe and sound back in Kentucky.[7]

David Brownlee from Washington County, Pennsylvania, settled in Trumbull County, Ohio, circa 1806. "In emigrating he brought with him a sow and a half a dozen pigs, five or six months old. They all seemed satisfied with their new Buckeye home, regardless of dangers from the prowling wolf, the bear, the panther, and the other wild beasts, plenty in our forests in those days, and lovers of pork, and indulged in it at every opportunity. These swine were in their sty every evening, and regularly at their troughs at feeding times, and things for a time went on very pleasantly with the porker family. Anticipation ran high with Mr. Brownlee in prospect of the good and profitable things coming in the shape of hams, shoulders, flitch, spare ribs, sausages, etc. Now one evening in early summer the pigsty was empty; none of its occupants put in an appearance. Not much solicitude was felt about their absence for a few days, then a diligent search was made for their whereabouts, but they could not be found and were given up for lost.

"After a time, Mr. Brownlee went back to Washington County to harvest the wheat that he had left growing. To his great surprise he found all his swine, with an addition of eight or ten pigs to the family, not one missing. When Mr. Brownlee was ready to return to his home he gathered his herd of swine, notified them of his purpose, and started them on their way. None making any determined opposition, they passed on before him until they came to the river, where they took to the water cheerfully and landed safely on the other side and took the direct road to Coitsville, nor ceased their efforts at all reasonable hours until they reached their Coitsville home and rested again within the sty, and fed from the trough which they had clandestinely deserted a few months before."[8]

And then there was Jezebel. Jezebel was a pig with a mind of her own. Her story was told by Warren Everhart, a county recorder during the 1950s who regularly contributed articles to a local newspaper about early pioneers in the area.

"Bears and panthers could be lived with without too much discomfort if the pioneer just gave a little practical attention to his poultry and flocks of sheep, which were about the only things that would be attacked for food. But in a way the pioneer was the author of his own undoing. Being a meat eater himself he had no sooner gotten well settled until he turned some of his razorback hogs loose in the woods near his home. Now, razorback hogs were smart animals and after they have run loose in the woods for about two years, they became the smartest, the most independent, cantankerous and dangerous wild beasts that ever roamed Dugan prairie or the hills that bordered Mad River Valley.

"When Ab Barrett came from Kentucky, one of the animals he brought with him was a razorback sow he called Jezebel. Ab settled near Buck Creek and for a few weeks he kept Jezebel in a rail pen but after a time he got tired of feeding her and turned her loose to fend for herself. All went well for awhile, then Jezebel disappeared. Two weeks later, a letter from Ab's brother in Kentucky stated that Jezebel was back in Kentucky. 'There is no mistake about it,' the brother wrote, 'your ownership mark is plain as day in old Jez's ear. She came back here on the hunt of a boyfriend and since that difficulty has been taken care of, you can come and get her again.'

"Well, Ab brought the sow back to Champaign County and this time he confined her in a lot surrounded by a good rail fence. Jez had no intention of raising a family under such confined conditions and, day by day, she kept scratching her back at a particular corner of the rail fence and, day by day, the corner where the rails joined spread little by little.

"As Jezebel's time approached all she had to do was give that fence corner a good hist with her tough old snout and she walked through. Ab, of course, knew why Jez had broken prison because from the time of a razorback's first creation, the female of the species always has been wont to hide away in some secluded spot and there, beside a fallen tree or some like object, she gathers weeds, twigs, leaves, etc. with her mouth, and piles them by the log in great heaps, sometimes three or four feet high. Then crawling beneath this great pile, the sow's family of little piglets are born.

"Ab waited a few days, and thinking the sow would be getting hungry he decided to look her up. He proceeded through the forest with caution for, from past experience, he well knew that a razorback sow with a newborn litter of pigs was the most dangerous beast in the forest. It made no difference to that sow how many times Ab had fed and slopped her, any man or animal that walked on legs was now an enemy, out to steal one of her babies and in order to protect them she had no fear of any living thing.

"A stream of spring water issued from the foot of a hill near Abner's home and then coursed its way toward Buck Creek. Ab guessed that the sow would have her nest near this stream because razorback intuition taught them to farrow where they could easily get water. Following the little stream, he spied old Jez's nest ahead of him. He stopped immediately for he had no intention of going any closer. He looked carefully in all directions but he could see no sign of the sow so he knew she very likely was in the nest with her piglets. Her nest was about a hundred feet from the water, so tiptoeing quietly so as to make little noise, he followed the stream to where she had left her tracks at the stream when she came for water. There he left a few feeds of corn, and re-tracking his steps as carefully as he had come, he went back to his house.

"Ab came every few days to bring feed to the sow, several times seeing her out of the nest but never far from it. Once she was eating some kind of small animal. Ab guessed an opossum had ventured too close. Razorbacks were meat eaters, they would kill and eat anything, even rattlesnakes.

"It was three weeks before Ab saw any of the piglets and he did not get very close. He was a good two hundred feet away when old Jez with her bristly hair standing up straight, started toward him. Ab backed away in a hurry and the sow stopped. He ventured no closer but dropped his feed and retreated. Next time he came, the feed had been eaten. Neither Jez nor the little pigs were in sight.

"Although food was available for the sow and her pigs, the plump little baby pigs were themselves a juicy treat for foxes, wolves, and panthers, as well as large members of the hawk family.

"A couple of times, thinking that old Jez was just an ordinary hog, foxes were the first to test the razorback's ire. The foxes, instead of feasting on a juicy pig, themselves became meat for the sow and her brood.

"The panther fared a little better, he managed to loosen himself from the grip of old Jez's jaws and limped away with a badly chewed-up hind leg. But the beast that really got the surprise of his life was a bear.

"A bear in his right mind would not pay any attention to a sow and her litter. Old Jez knew this and she therefore went her way, and the bears did the same, except for this one bear.

"This was a most curious bear and while wandering around a pioneer's distillery one night, he sampled the distiller's mash. He became so tipsy that he imagined himself the monarch of all he surveyed. In this condition it was his misfortune to meet old Jez. Rearing up on his hind legs, the bear beat his chest with his forepaws and let out a mighty roar. Jezebel accepted the challenge and charged right in. She well knew the force of a bear's strike with his forepaws, but the sow ducked low and the bear missed. Jez bit his hind leg broadside. The drunken bear went down in a heap and, before he could recover, Old Jez had him by the hind leg and was doing her best to rip it into shreds. She hung on for at least fifty feet before the bear got loose. By that time, he was sobered-up.

"Jezebel found she could not exist on worms, frogs and snakes alone, so two or three time a week she led her brood back to Ab Barrett's farm and there in her old rail pen she and her babies always found a generous helping of Ab's corn. In this manner Ab and the razorback became good friends again and the pigs learned not to fear him.

"One of the pigs was an exceptionally large and growthy male, half again as large as any of the others. Ab and some of his neighbors decided to save this fellow for a razorback sire.

Ab named him Samson, after the Biblical character of old, but Samson's history as a razorback boar is a story in itself. I may write it."[9]

This one story proved to be the only tale, however, and nothing more was heard either of Jezebel or Samson.

Mrs. Sarah Rockel told how her family, in 1845, "lost a sow in the early spring and late the next fall when the snow flew, a drove of fine porkers came grunting to the barnyard gate. The old sow had littered, reared her brood on mast and with the coming of ice and cold remembered her slops and straw."[10]

Many of the early pioneers constructed solid pens of logs to confine their pigs and to protect them from predators. Other families gave only minimal attention to feeding or caring for their livestock, allowing them to run free in the surrounding forest and fend for themselves. These hogs took shelter wherever they could find it. Unattended for weeks and months at a time, they often intermingled with the wild hogs already ensconced in the Ohio wilderness.

Mary Gray Gutheridge wrote in her memoirs about the deplorable condition of the cabin her father purchased in 1815. "When we came in sight of Mingo Valley, it was a beautiful sight" But "when we got to the cabin—such a place to live!" The wild hogs had taken possession of it. . . . "We sat down and cried. . . . My oldest sister began laying down the puncheons the wild hogs had rooted up. A neighbor came . . . and helped us fix the puncheon floor."[11]

Charles A. Wiant, as a small boy, attended school in a one-room, log building that was elevated off the ground by large boulders, one at each of the four corners. The hogs that ran loose in the neighborhood often sought shelter under the school where their loud grunting occasionally became annoying. When they scratched their backs by rubbing against the underside of the floorboards, Charlie remembered how the older boys, all of whom carried pocketknives, would open them and stick the blades down through the wide cracks into the pigs' backs to make them squeal.[12]

Daniel Snell came to Darke County with his parents in 1809. During their first winter in Ohio they cordoned off a corner of their eighteen by twenty foot, dirt floor, log cabin for an old sow and her pigs to keep them from freezing.[13]

Virtually all farm families raised at least one or two hogs in order to provide pork for their own consumption. Salt, which was used for preserving meat, was such a scarce commodity that some people, as an alternative, resorted to burying the whole carcass deep in the snow during the cold winter months and cutting off pieces as they needed it.

William Owens not only is considered the first white man to settle and live on land south of Westville, Ohio, in 1779, but also is believed to have been the first to introduce domesticated hogs to the area. When Owens arrived and chose property overlooking the Mad River Valley, he formed a friendly relationship with the Indians who lived along the banks of the Mad River. Because he kept hogs, which the Indians at that point in time found most unusual, the tribe referred to him as Koske Elene, or "Hogman." Although the Indians were unaccustomed to eating pork, Owens eventually persuaded them to taste it. A county history reports that they became so fond of the taste they would trade a whole deer for just one slab of bacon.[14]

John Butler, a justice of the peace in Franklin township, Wayne County, raised hogs. An Indian shot one of them and carried it off. Mr. Butler followed the culprit to his village and demanded the thief pay him for his loss. Asked why he'd killed the hog, the Indian replied, "I wanted grease." Made to pay two deer skins, the disgruntled Indian sullenly kicked them toward the white man. A short time afterwards, the Butler's log cabin was burned to the ground.[15]

In Darke County, a Mr. Wyland played a contemptible trick and lost his only sow as a result. Two men, one of whose name was Connor, learned that the Wylands were asking for help to erect a cabin and offered their services. The cabin was raised and the two men decided to camp for the night in the woods adjacent to the cabin. They built a fire, ate their supper, shared the contents of a jug of whiskey, propped their rifles against a tree and, because it was cold and beginning to snow, hung their moccasins on sticks near the fire to dry. While they slept, Wyland crept to the campsite where he drank some of their whisky and poured the rest on the ground. He tossed their moccasins into the fire and returned to his cabin. Watching them next morning, he saw the men discover the empty jug and, in disgust, throw it down. When they realized their moccasins were in the fire, both looked down at the tracks in the snow which quite plainly led to the cabin. Wyland grew uneasy. After a last menacing glance, one of the men started off barefooted in the snow toward the settlement. Connor, however, pointedly strode to the pig pen. There was a loud rifle shot. A few minutes later, prominently displaying something black in his hand, he abruptly turned and, he, too, walked barefooted in the direction of the settlement. Apprehensively, Wyland hurried to the pig pen where he found his young pigs unharmed but his sow lying dead and partially skinned.

Several weeks later both Wyland and Connor attended a camp meeting. When the two men saw each other, Connor deliberately positioned himself so that, stretching out his long legs, Wyland would be aware of his new moccasins, moccasins made from the hide of a black sow.[16]

David Tod served from 1862 until 1864 as Ohio's twenty-fifth governor. As a young boy, "the spirit of humor overflowed with him" and once, along with a friend, he pulled a prank on a neighbor. The neighbor was himself a mischievous soul who enlivened the vicinity with his harmless antics. The boys knew that their victim fed his hogs in the same spot each evening so David sawed the top rail of the fence nearly in two. When the neighbor leaned over the fence to pour feed into the trough, the rail broke and man, swill, and fence rail fell into the pen. Hearing the boys' suppressed laughter, the man surmised how the "accident" had happened "but he climbed from the mess and said nothing."[17]

Reverend Dyer of Adams County, in 1820, was riding at a fast trot along the trail toward his home when his horse suddenly shied at a black hog rooting under a tree. After reaching home, the Reverend commissioned life-like pictures of black hogs painted on the side of his barn "where," he said, "the horse could shy at them at her pleasure."[18]

In 1831, a Baptist minister by the name of William Miller established a new denomination known as "The Second Adventists," or, more commonly, "Millerites." He predicted the end of the world on a certain date and some of his converts immediately began selling their property "to get the free use of cash for the short time they felt they were to stay here below. One of these went to a neighbor to sell a young pig. The latter demurred. 'Too young.' 'No,' rejoined the Millerite, 'he'll grow.' 'Not much, for according to your belief, he will be roasted pig altogether too soon for my use.' "[19]

By eastern standards, Ohioans were provincials. A traveler through the territory in 1802 called it a "backwoods" and wrote, "Their Generals distill whiskey, their Colonels keep tavern and their statesmen feed pigs."[20] Another said, ". . . pigs . . . swarm in this part of the country like grains of sand on the seashore. . . ."[21]

Even in towns, travelers were quick to point out that one of the most common sights was the abundance of pigs running free along the streets. Two brothers from England, Daniel and

William Constable, "were amused at the hogs running in the streets and for their speed which compared to greyhounds."[22]

Cyrus P. Bradley, who traveled down the Ohio River in 1835, stopping at towns along the way, wrote in his journal about seeing pigs wallowing everywhere in the streets and gutters. In Cincinnati, he read an account in a morning paper that "one of these ravenous beasts seized a young child by the arm, tore him from his mother's doorstep into the gutter, where, had it not been for the child's screams and the interference of a gentleman, he inevitably would have devoured the child. This was a little too bold."[23]

In his book, *Historical Collections of Ohio*, Henry Howe talks about a visit to Eaton, Ohio. ". . . the landlady at the village tavern was a comical, good-natured creature, whom, if I rightly remember, the young men of the village (who largely boarded with her) addressed as 'Aunt Sally.'

"In those days the pigs had the liberty of the streets in the small towns; yes, even in Cincinnati they roamed abroad, doing good scavenger work, while sending forth their notes loud and strong.

"Whether Aunt Sally was unwedded or wedded I know not, but she evidently felt the want of some object to pet. Woman's heart has many tendrils and sometimes these fasten queerly; hence, Aunt Sally's especial attentions to a pig, which were gratefully returned, all to the daily amusement of her boarders.

"Piggie was not over cleanly, had only one ear, some dog having appropriated the other, and once, to my astonished eyes, during my stay, dashed into and through the house with the freedom of one of the family. I was told he had once even appeared in the dining room."[24]

At Upper Sandusky, a Bostonian who was spending the night at an inn became so annoyed at another guest's snoring he retreated to the stagecoach to sleep. It proved to be an unwise move for "hogs grunted round the stagecoach so hideously, he was afraid to come out again and lay there shivering till morning."[25]

As settlements turned into villages and towns, and the population increased, citizens struggled for better living conditions. Columbus, Ohio, appointed a pig catcher to round-up free-roaming hogs and confine them in pig "pounds." At a June, 1800, court session in Adams County, "The Court allowed Samuel Pettit three shillings and six pence per panel for getting, hauling, and putting up twenty-four panels of post and railing for a stray pen."[26] And on March 27, 1828, in Urbana, Ohio, the "City Fathers ordained that there shall be levied a tax of 12 1/2¢ on all hogs owned in the town of Urbana and running at large in the public streets thereof, provided that all hogs kept within enclosures in said town so as not to be a nuisance, shall be exempt from said tax." On May 9th of the same year, "The ordinance was amended doubling the tax for hogs using the streets for pasture, making it a quarter for each one running at large."[27]

In St. Paris, Ohio, Mary Harris, better known as "Aunt Mary," a one-time slave in Virginia, kept hogs in her back yard. Nearly every day she could be seen walking the streets in town collecting garbage to feed them. An item in a St. Paris newspaper once noted that she was fined $1.00, not because her hogs were running loose but, for having too much trash around her house. In 1883, one of her sows escaped and she advertised: "A black sow strayed away last Friday. She had a white streak running from up her shoulder and a white spot in her forehead. Five dollars will be paid for her return to the undersigned. If taken up and sold, let me know it; they will not lose by giving the information. Mary Harris, Nov. 3, 1883."[28]

In 1878, a black man, Washington Staunton, owned a few pigs and a couple of horses, all of which he permitted to run-at-large during the nighttime. Morning after morning, Erskine Bemis, a neighbor, found the animals on his property and one day, exasperated beyond endurance, penned them in a lot. Staunton appeared and matter-of-factly proceeded to tear down the fence. An altercation between the two men ensued, terminating when Staunton struck Bemis with a brick. Staunton was arrested and fined $1.00 plus costs for allowing his stock to run at-large. No mention was made of the fight.[29]

P. B. Ross, irritated at a neighbor, advertised in his local paper: "Persons who have been turning their stock in the pasture of P. B. Ross, without consent, are notified that unless they stop the practice, they will be prosecuted."[30]

In the rural areas of Ohio, much of the livestock ran loose and proper identification was a problematic issue. It became imperative to mark animals in such a way that everyone could claim his own stock and not someone else's. Legislation was enacted as early as 1796 in present-day Cuyahoga County requiring every livestock owner to register his special mark with the township clerk. Most families used the same marks for their hogs and their cattle. Samuel Ewing used "a crop of the left ear and a slit and under bit in the same."[31] Another early settler "pointed one ear and cut a V-shaped piece out of the other."[32]

William M. Roberts inserted a notice in an 1814 paper that a hog had strayed onto his property: "Came to my house . . . sometime in April last, a large barrow hog, marked with a swallow fork and under bit in the right ear and a crop off the left, which hog I killed, after fattening him with mine, and he weighed two hundred and thirty-five and three fourths pounds. The owner may come and prove property, pay charges and receive pay for the same."[33]

Although earmarks provided some legal recourse for livestock owners, once the animals were sold out of the neighborhood and butchered, ownership could not be proven.

Altering a mark was illegal and anyone convicted of it was liable for a $5.00 fine.

Not everyone marked their livestock. John Dilts, for example, did not mark his animals for he used a visual description when inquiring for his lost hog. "Strayed from the subscriber, Nov. 12, 1869 . . . near Pollocktown, a black and white (more black than white) sow, with long tail, ears hanging somewhat over the eyes, would weigh 180 or 200 pounds. Any person giving information of her whereabouts, or returning her to me, will be rewarded for their trouble."[34]

In 1790, during the second court session in the Northwest Territory, a man was tried for stealing a hog: "A grand jury was impaneled and found a bill against Ezra Lunt for stealing a hogg."[35]

In September of 1858, Frederick Gessler of Westville, Ohio, was arrested and jailed. The charge was hog stealing. "He had been doing an extensive business in the pork trade during the summer," read the account.[36]

One day in 1803, shortly before a court session in Greene County commenced, a man by the name of Owen Davis charged a certain individual with stealing his hogs. The accused man angrily denied the allegation and the two began fighting. Davis, who emerged the victor, marched into the court room and addressing the judge, said, "Well, Ben, I've whipped that d___n hog thief—what's to pay? " He threw a buckskin purse containing eight or ten dollars onto the table and added, "Yes, Ben, and if you'd steal a hog, d___n you, I'd whip you too."[37]

Two Logan County neighbors disagreed over the ownership of a hog. Both claimed the animal. The defendant, an unscrupulous sort of fellow, persuaded a friend to swear before a Justice of the Peace that he knew the pig had been raised by the defendant. Later, it was proven that the defendant had "raised" the pig by stooping over a low fence at the plaintiff's barn and "lifting it" out of the pen.[38]

At a circa 1817 spring election in Logan County, two men were selected to serve as fence viewers. Fence viewers, those persons whose duty it was to examine the split rail fences in case of disputes, were an essential part of everyday life in early Ohio. According to the law, a fence had to be in sufficient condition to keep an owner's stock in and other livestock out, before a property owner could recover damages caused by someone else's animals. The 16 1/2 foot long "perch pole," the length of one rod, served as the fence viewers' unofficial badge of office. Ordinarily, only one man was elected; however, the voters in Logan County had a droll sense of humor that particular year and chose two men. Adam Rhodes, a remarkably tall man, was chosen to "chin the fence." In like matter, Hiram White, who was an unusually short man, was appointed to check the "pig holes," the openings near the ground.[39]

Lorenzo Dow, one of the notable evangelistic ministers on the frontier, was responsible for apprehending a hog thief. Dow, possibly the most theatrical itinerate circuit rider in this part of the country, fascinated crowds for over thirty years with his unconventional mannerisms and attire. A small man with long flowing hair parted in the middle and a long, unkempt beard, he customarily wore a straw hat and a white blanket overcoat. He would arrive at the appointed meeting place, pull off his hat and coat and toss them onto the ground. He began each sermon with "Hell and damnation," followed with a string of oaths "enough to frighten the wickedest man." Then, beseeching the congregation to repent their sins he would launch into a lengthy and solemn discourse.[40]

Prior to one of his memorable sermons, he was informed that someone in the neighborhood was stealing their pigs. Without a moment's hesitation, he stooped and picked up a rock. Stating in a commanding voice that he would reveal the thief by throwing a rock at him, he made the motions of throwing a rock. A man in the audience dodged. "There's your thief," thundered Dow. The man admitted his guilt.[41]

Relatively early in the time-frame of Ohio history, farmers made efforts to improve the quality of their swine. In the forefront of the movement was a religious group known as the Shakers. The Shakers, progressive agriculturists, had established a community in Warren County, and about 1816 started a crossbreeding program using a Big China boar and several brood sows they imported from the eastern states. These animals, bred to the Shakers' domestic hogs, resulted in an animal that was widely acclaimed, paving the way for the creation of a new strain labeled "Poland China." A monument, a rectangular, granite block with a simple bronze plaque, outside the small village of Blue Ball, Ohio, marks the general vicinity where the "birth" of the Poland China hog took place.

In an 1850 agricultural report, farmers in the state who responded to a questionnaire listed the kinds of hogs most often found in their communities. A few of the names included were Byfield, Irish Grazier, Calcutta, Russian, and Big China —no mention yet of Poland Chinas for the name as a separate breed had not yet been drafted.[42]

Throughout most of Ohio's embryonic years, pork was accepted in lieu of cash by nearly every merchant. Pork, usually in the form of hams and bacon, was used in exchange for goods of all sorts—even land. In August of 1801, John McDougal purchased two hundred acres in Ross County for $2.00 an acre. Payment was made partially in cash and partially in swine. A 1,000 acre tract of land in the same area sold for a combination of cash, horses, cattle, and swine.[43]

Ohio quickly became an important center for hog production and, like cattle, they were driven on foot to the marketplace. Preparations for the drives began in September when drovers and dealers collected the hogs for which prior arrangements had been made. Each animal was caught and a noose slipped over its mouth to keep it from biting while its ears were stitched together to cover its eyes. Some drovers preferred stitching the eyelids shut so the animal, in essence, was blind. Because the unruly animals were unaccustomed to being herded, much less for any great distance, the procedure was considered a practical one. Pigs were creatures who took fright at every sight and sound and would scatter in all directions. They sometimes made their way back to their place of origin which would necessitate collecting them over again, much to the dismay of the drovers.

Weighing stations frequently were set-up in the center of a town and it was a familiar sight to see hundreds of hogs penned there preceding a drive. Each pig, squealing in protest, was lifted in a leather sling suspended from a steelyard and individually weighed. Its weight was recorded and a patch of bristles cut from the hindquarters to indicate that the pig had been weighed. A crowd of townspeople usually gathered to watch.

Droves of 1,000 to 1,200 hogs were not uncommon. Neither was it unusual to see these large herds following an equally large herd of cattle for by trailing after cattle, hogs could forage on the wasted feed. Flocks of sheep, geese, ducks, and turkeys sometimes completed the drove! After two or three days on the road, pigs became more manageable and could average a gait of eight to ten miles a day.

Before a drove was permitted to pass along a toll road, fees were collected. On occasion, a count was recorded and the drover was allowed to pay the amount on the return trip. In 1809, a general charge was 10¢ for every "score of hogs," a "score" being a unit of twenty. A sign posted on a toll road between Westville and St. Paris, Ohio, listed pigs at 1/2¢ each.

After an arduous journey of from four to six weeks to far distant cities, dealers often found an uncertain market. More than one person met financial ruin.

Thomas Moore, of Clark County, assembled a drove of hogs which he took to Baltimore, contracting to pay the owners for them afterwards. In place of money, one of the women requested Moore bring her a length of silk fabric and the latest, most fashionable pattern. He agreed. The venture was a disaster for slaughter house prices were poor and Moore's expenses substantially more than he had anticipated. His losses were heavy. The lady received her fabric and pattern but no one else was paid at the time. The remaining owners, however, were promised full restitution and although it took him several years to accumulate sufficient funds, Thomas Moore made good his promise and paid every penny he owed.[44]

When Edward Popejoy, from Fayette County, learned that hogs customarily were worth from $1.00 to $1.50 per hundred weight, he decided to try his hand and take a herd of hogs to the East. He made the rounds of his neighbors buying those pigs he thought were heavy enough to be sold. When he asked one of the farmers how much he wanted for his hogs, instead of stating a price, inquired of Popejoy how much he was willing to pay. Popejoy named his price, whereupon the neighbor said, "That's far too much," and suggested a figure he considered more fair. Popejoy accepted.[45]

Although there were many long drives to Baltimore and Philadelphia and shorter drives to Sandusky, Dayton, and Columbus, Ohio, the largest market in Ohio was in Cincinnati. By 1845, Cincinnati was the hog packing center of the country. It was said that packers in this riverside city "originated and perfected the system that packed fifteen bushels of corn into a pig and then packed that pig into a barrel and sent him over the rivers and oceans to feed mankind."[46]

Cincinnati, dubbed "Porkopolis," slaughtered 400,000 hogs in 1849 alone.[47]

Cincinnati, in 1811, "a little town under the hill,"[48] grew to become one of the great cities along the Ohio River. There was constant loading and unloading of the boat traffic. Almost from the beginning of its settlement in 1788, Cincinnati was a commercial center. Stores carried merchandise found in few other places in the state. It was adjudged the cultural capital of the New West and "Queen of the West." Nonetheless, not everybody described it in a favorable light. One 19th century visitor said ". . . I should have liked Cincinnati much better if the people had not dealt so very largely in hogs."[49] "A caustic English traveler described the air as polluted with the odors of the rendering plants, the brooks reddened with the blood from the slaughter houses, and the grassy spaces strewn with pigs' tails and jaw bones. . . ."[50]

Nothing about a hog appeared to go to waste. The blood, hair, and hoofs were sold to manufacture potash; the offal was used for fertilizer; the fat went into lard or was converted into candles and soap. Even the clean, dry, hog bristles sold for 62 1/2¢ per pound.

Sausage was referred to as "pulverized hog," by many Cincinnati townspeople but the gentry considered the expression vulgar and uncouth.

A few Ohioans sought more southern markets to sell their hogs. S. A. Butterfield, of Montgomery County, for instance, made frequent trips to New Orleans via a flatboat where, if the price was not high enough, booked his shipment on to Havana, Cuba. He was involved in a ship wreck on one of his trips to Havana but reacting quickly, he opened the pens, which were on deck, shoved his pigs overboard, and they all swam to shore. Except for a small number which were stolen, Butterfield rounded-up the rest, sold them profitably, and returned home eager to try it again.[51]

Breeders in the northern part of Ohio frequently drove their hogs to Detroit where they were slaughtered and shipped to Montreal, Canada, by way of the Great Lakes. After 1825 when the Erie Canal opened, carcasses were shipped to New York City. The boats made stops in Buffalo and other cities along the way where transactions often took place. Women flocked to the wharf when a shipment arrived. They would inquire of the ship captains if there were barrels of lard aboard. If the answer was "yes," they would purchase it on the spot.

Although women often used copious amounts of lard in their cooking, preparing pork did not necessarily call for its use. Boiling was an option. "To Boil A Leg of Pork" was one of the featured recipes in *The Compleat American Housewife 1776*. It read: "A leg of pork must lie in salt six or seven days after which put it into a pot to be boiled, without using any means to freshen it. It requires much water to swim in over the fire, and also to be fully boiled; so that care should be taken that the fire does not slacken while it is dressing. Serve it with melted butter, mustard, buttered turnips, carrots or greens."[52]

Towns throughout Ohio held weekly markets where farmers in the community offered their produce for sale. In addition to firewood, hay and straw, and the fruit and vegetables from their gardens, people from the countryside sold all kinds of animals. This included live pigs, calves and poultry. Because driving hogs to market along a particular path in Miami County became a commonplace sight, local residents called it "Hogpath Pike." Today, the heavily traveled road between Troy and Pleasant Hill has been designated St. Rt. 718 but there are those in the community who still occasionally refer to it as "Hogpath Pike."[53]

In the summer of 1832, when the temperance movement first began, a gentleman in Clark County attended one of the meetings and was inspired to sign a pledge agreeing neither to consume nor to keep any alcoholic beverages in his home. Because it was the accepted custom at that time to provide laborers with whiskey and the man who had signed the pledge was a farmer and on occasion hired extra men, there was some concern his help might leave. None did, however. They obligingly volunteered to assist in dumping the whiskey and the cherry-bounce which was stored in the cellar of the house. They inadvertently poured it all into a little gully that ran through a lot where market hogs were penned. Convulsed with laughter, a son told the story many years after the event that every one of the hogs became intoxicated, uncontrollably staggering and reeling on unsteady legs. He said, "I never see a drunk man that I do not think of those hogs."[54]

In spite of the temperance movement, intoxicating beverages continued to play a salient part in the lives of many Americans. Vessels of amazing sorts were utilized to carry the fermented drinks. Even hog skins were called into duty. An article elaborated, "The skin is stripped from the hog almost intact, turned inside out, and then with the legs, tail and throat tied up it is filled." An unsuspecting traveler described his introduction to this type of container. "The baggage car of our train was nearly filled with these uncouth objects and until we knew what they were we supposed it was a load of hogs."[55]

Hogs played a significant role in feeding America's armies. Herds of hogs as well as cattle often were driven in the wake of military units in order to provide fresh meat for the men. In the winter of 1813, Major Samuel Myers of Fayette County was employed by army contractors to superintend the transportation of eight hundred hogs to Fort Wayne, Indiana, where General Anthony Wayne, commander-in-chief of the Northwest Army, was stationed. It was the responsibility of Major Myers to see they arrived safely in Fort Wayne. With a guard of twelve soldiers, the eight hundred hogs, a number of cattle, and about forty packhorses and a few assistants, the group started through the thick forest to Fort Wayne, "occasionally stopping to allow the hogs to feed on the abundance of acorns in the forest. . . . Although Indians were plenty they passed on quietly."[56]

When the Confederate General, John Morgan, and his band of raiders, during the Civil War, rode through southern and eastern Ohio and one day crossed the Muskingum River near McConnelsville, some in the neighborhood took-up arms while others hid. As news of the troop's approach reached a young farmer, he "sought safety in a pig pen and laid down, as he thought where he could not be seen, crouched behind a matronly specimen who was attending to the gastronomic requirements of a new-born progeny. He had been seen to flee by one of the troopers, who, on coming to the pen, looked in and espying the poor frightened fellow, exclaimed with a grin; 'Halloa! how did you get here? Did you all come in the same litter?' "[57]

While fighting the British during the War of 1812, a group of seven thousand Kentucky volunteers came to Ohio and marched toward Detroit where the enemy was entrenched. Along with the seven thousand men came one small pig.

As the forces were leaving Harrodsburg, Kentucky, the men noticed two little pigs by the side of the road. One began following them. All day long it kept pace and that evening the men shared their meal with it. In the morning, the men again shared their food. The pig continued walking with the troops and while they crossed the Ohio River on ferries, it swam across. After that notable feat, the men decided to adopt the pig for their mascot. It marched with them as they tramped northward through Ohio but to insure its safety while the soldiers engaged in battle, the little fellow was left in the care of a family near Lake Erie. Upon the troops' return, the pig again accompanied them back through Ohio, across the Ohio River and home to Kentucky. He was acclaimed "The Pig

Who Went to War" and was given a hero's welcome. His reward was a home for the rest of his life on the farm of the Governor of Kentucky.[58]

Chapter 9: PIGS IN OHIO

1. Joshua Antrim, *History of Champaign and Logan Counties, From Their First Settlement* (Bellefontaine, Ohio: Press Printing Co., 1872), p 28.
2. W. H Beers, *History of Montgomery County* (Chicago: W. H Beers & Co., 1882), p 294.
3. Judge Evan P. Middleton, *History of Champaign County, Ohio* (Indianapolis, IN: Bowan & Co., 1917), p 347.
4. Joshua Antrim, *History of Champaign and Logan Counties, From Their First Settlement* (Bellefontaine, Ohio: Press Printing Co., 1872), p 238.
5. Henry Howe, *Historical Collections of Ohio* vol. 2 (Cincinnati, Ohio: Krehbiel & Co., 1904), p 209.
6. *Ohio Historical Review* (Columbus, Ohio: pub./ed. Ralph S. Estes, n. d.),p 24.
7. W. H. Beers, *History of Champaign County, Ohio* (Chicago: W. H. Beers & Co.,1881), p 489.
8. Henry Howe, *Historical Collections of Ohio*, vol. 2 (Cincinnati, Ohio: Krehbiel & Co., 1904), p 614.
9. Warren Everhart, "Jezebel" (n. p., c 1956).
10. Orton G. Rust, *History of West Central Ohio* (Indianapolis, IN: Historical Pub. Co., 1934), p 353.
11. Warren Everhart, "Story of Mary Gray Gutheridge" (n. p., c 1956).
12. Charles A. Wiant, the author's maternal grandfather.
13. W. H. Beers, *History of Darke County* (Chicago: W. H. Beers &.Co., 1880), p 757.
14. ——, *History of Champaign County* (Chicago: W. H. Beers & Co., 1881), pp 213-404.
15. Henry Howe, *Historical Collections of Ohio,* vol. 2 (Cincinnati, Ohio: Krehbiel & Co., 1896), p 835.
16. W. H. Beers, *History of Darke County* (Chicago: W. H. Beers & Co., 1880), p 414.
17. Henry Howe, *Historical Collections of Ohio*, vol. 2 (Chicago: Krehbiel & Co., 1904), 182-183.
18. Nelson Evans, *History of Adams County* (West Union, Ohio: Emmons Stivers, 1900), p 529.
19. Henry Howe, *Historical Collections of Ohio*, vol. 2 (Cincinnati, Ohio: Krehbiel & Co., 1904), p 49.
20. W. H. Beers, *History of Brown County* (Chicago: W. H. Beers & Co., 1883), p 264.

21. *The Ohio Frontier: An Anthology of Early Writings*, ed Emily Foster (Lexington, KY: University of Kentucky Press, 1996), p 123.

22. J. Brian Jenkins, *Citizen Daniel (1775-1835) and The Call of America* (Hartford, CT: Aardvark Editorial Services, 1997), p 31.

23. Cyrus P. Bradley, "*Journal of Cyrus P. Bradley, A Trip Through Ohio and Michigan in 1835*" (Columbus, Ohio: Ohio Archaeological and Historical Quarterly, 1906), p 221.

24. Henry Howe, *Historical Collections of Ohio*, vol. 2 (Cincinnati, Ohio: Krehbiel & Co., 1904), p 448.

25. *The Ohio Frontier: An Anthology of Early Writings*, ed Emily Foster (Lexington, KY: University of Kentucky Press),p 117.

26. Nelson Evans, *History of Adams County* (West Union, Ohio: Emmons Stivers, 1900), p 93.

27. Clipping from a 1930 *Urbana (Ohio) Daily Citizen*.

28. *St. Paris (Ohio) Dispatch*, November 9, 1883.

29. *Urbana (Ohio) Daily Citizen and Gazette*, May 16, 1878.

30. ———, November 27, 1889.

31. W. H. Beers, *History of Madison County* (Chicago: W. H. Beers & Co., 1883), p 651.

32. Henry Howe, *Historical Collections of Ohio*, vol. 2 (Cincinnati, Ohio: Krehbiel & Co., 1904), p 736.

33. *(Urbana, Ohio) Spirit of Liberty*, Feb. 8, 1814.

34. *Urbana (Ohio) Citizen and Gazette*, December 27, 1869.

35. Walter Havighurst, *River To The West* (New York, N.Y.: Penquin Group, Inc., 1970), p 90.

36. Name uncertain, possibly *Ohio State Democrat*, September 10, 1858.

37. W. H. Beers, *History of Champaign County* (Chicago: W. H. Beers & Co., 1881), p 209.

38. Joshua Antrim, *History of Champaign and Logan Counties From Their First Settlement* (Bellefontaine, Ohio: Press Printing Co., 1872), p 238.

39. ———, p 237.

40. ———, p 161.

41. Henry Howe, *Historical Collections of Ohio*, vol. 1 (Norwalk, Ohio: Laning Printing Co., 1896), p 414.

42. *Report of The Commissioner of Patents For The Year 1850* (Washington D.C.: Office of Printers to House of Representatives, 1851), p 397.

43. R. Douglas Hurt, *The Ohio Frontier* (Bloomington & Indianapolis IN: Indiana University Press, 1996), p 109.

44. W. H. Beers, *History of Clark County* (Chicago: W. H. Beers & Co., 1881), p 384.

45. R. S. Dills, *History of Fayette County* (Evansville, IN: Unigraphic, 1881), p 1005.

46. *A Geography of Ohio*, ed Leonard Peaceful (Kent, Ohio: Kent State University Press, 1996), p 254.

47. R. E. Banta, *The Ohio* (New York, N.Y.: Rinehart & Co., 1949), p 448.

48. Joshua Antrim, *History of Champaign and Logan Counties From Their Earliest Settlement* (Bellefontaine, Ohio: Press Printing Co., 1872), p 31.

49. *The Ohio Frontier: An Anthology of Early Writings*, ed Emily Foster (Lexington, KY: University of Kentucky Press, 1996), p 168.

50. Simeon Fees, *Digest of Ohio History* (Chicago: Lewis Pub. Co., 1937), p 317.

51. W. H. Beers, *History of Montgomery County* (Chicago: W. H. Beers & Co., 1882), p 448.

52. Julianne Belote, *Compleat American Housewife, 1776* (Conrad, CA: Nitty Gritty Productions, 1974), p 72.

53. *Miami County History,* Miami County Ohio Sesquicentennial Committee (Columbus, Ohio: F. J. Heer Printing Co., 1953), p 290.

54. *Ohio Historical Review* (Columbus, Ohio: pub/ed Ralph S. Estes, n. d.) pp 17-18.

55. *(Urbana, Ohio) Monthly Visitor*, vol. 17, no.1, ed James Hearn, 1889.

56. R. S. Dills, *History of Fayette County* (Evansville, IN: Unigraphic, 1881), pp 290-291.

57. Henry Howe, *Historical Collections of Ohio*, vol. 2 (Cincinnati, Ohio: Krehbiel & Co., 1904), p 312.

58. Magazine clipping from *Country Kids* (n. d.), p 37.

THERE WERE SHEEP

Chapter 10

Although it remains uncertain when the first sheep were brought to Ohio, Seth Adams, who settled near Zanesville, is known to have had sheep in 1807. With twenty-five to thirty head, crosses between native sheep and the purebred Merino sheep he had imported at great expense from Spain, Adams left Dorchester, Massachusetts, for a tract of land on Wakatomaka Creek. The sheep were driven as far as Pittsburgh. From there they were shipped by boat to Marietta and then, again, driven overland for the final stage of the trip to their new home.[1]

Adams' flock increased. Many of the animals, which were of a much higher quality than the usual domestic sheep, were sold to other people in the area for seed stock. In 1811, Adams lost more than thirty head in the same night to marauding wolves, "an event that discouraged him."[2]

In spite of Adams' misfortune, Merino sheep rapidly gained in popularity, especially after 1810 when it became less involved to import them from Spain. Not only because of the fineness and the weight of the fleece but also because of their ability to walk long distances, breeders preferred the Merino. Breed associations were formed and Merinos became a favorite type of sheep to exhibit at agricultural fairs. The *Report of The Commissioner of Agriculture For The Year 1867* included a picture of a Merino ram, Judge Laurence, that was shown with considerable success by Curtis Kelsy of Sidney, Ohio.

JUDGE LAWRENCE

from *Report of The Commissioner of Agriculture For The Year 1867*

Raising sheep in the wilderness was a gamble at the best of times. Predators found them easy prey. Bears could kill them with a swift blow of one powerful paw. Although sheep have slim ankles and thick, muscular upper legs which enable them to move suddenly and quickly, the timid creatures sometimes tend to huddle together in a cluster when alarmed.

A man in Miami County built a strong rail pen abutting the wall of his cabin thinking it would help protect his five sheep. One night, however, the wolves' persistent howling frightened the sheep inside their sturdy pen and, bleating in terror, they struggled desperately to get out. Eventually they succeeded and, with the wolves in close pursuit, bolted in the direction of the creek where the first animal plunged headlong into the deep water. The other four sheep followed and, as a consequence, all five drowned.[3]

An Indian and his dog passed a cabin in Madison County one day when the dog discovered the settlers' sheep—

"meketha" in Shawnese—nibbling grass in a nearby field. Barking excitedly, it began chasing first one and then another as they fled. During the tumult the dog killed one of them. The settler was so enraged that he shot the dog. The Indian, equally enraged, vowed to kill the white man. Only through the calm intervention of Jonathan Alder, the white man who had been reared as an Indian, was the matter resolved peacefully.[4]

Dogs as well as wild animals accounted for large numbers of sheep kills. An Ohio *Annual Report of The Secretary of State,* for the year 1885, noted that of the nearly 3,700,000 sheep raised in the state, 29,006 had been killed by dogs.[5]

Most of the sheep in early Ohio were a nondescript type. Although the fleece was of a relatively poor grade, it was, nonetheless, an important source of wool for people along the frontier. As food, the flesh of sheep, or mutton, was said to have been of secondary importance only.

In the spring after the men had shorn the sheep, wives and daughters spent long hours sorting the fleece, scouring it, carding it, and then spinning and twisting the wool into long hanks of creamy-white yarn. The wool from black sheep ordinarily was spun separately. The women spent more hours sitting at their looms, weaving the yarn into fabric from which garments were made. Linsey-woolsey, woven from flax and wool, was scratchy but serviceable and warm.

Elnathan Kemper hired a neighbor woman to spin the wool from his sheep in 1815 and paid her 12 1/2¢ a day. It took her three full days. The total bill came to 37 1/2¢.[6]

In the early 1800s, one expected to shear about three pounds of wool from each average sized animal. The usual cost of keeping a sheep for a year amounted to approximately 50¢ per head.

By 1816, and probably much earlier in some parts of the state, woolen mills were in operation in Ohio. Women preferred the machine-carded wool because it not only relieved them of the burden of cleaning the fleece and spinning the yarn but the process also made a better yarn and, ultimately, a more desirable piece of fabric.

A Miami County history told about a young man during the War of 1812 who was notified on a Saturday evening that he was to report for duty on the following Monday. The household was in instant turmoil for winter was near at hand and his heavy clothing was not ready. According to Mrs. Jane Scott McKinney, the soldier's sister, "for twenty-four hours the family worked without rest. Calling the men to shear the sheep, black and white to make the proper mixture, for they had no time do dye the wool, the women spun the thread, wove the cloth, cut and made the clothes, and the soldier on the next day went to his regiment clad in the garments that had been the property of the sheep the day before."[7]

Paul Fearing was one of the provident men in the state who became interested in improving the quality of sheep. In 1808, he is reported to have paid eight hundred dollars for a ram and over three hundred dollars apiece for a number of ewes—an enormous sum at that time.[8]

Thomas Worthington, who served as Ohio governor from 1814 to 1818 and has been called the "Father of Ohio Statehood" because of his leadership in the statehood movement, raised sheep on his farm at Chillicothe and constructed his own woolen mill. An account book shows that he paid two hundred fifty dollars each for a purebred Merino ram and several ewes.

Although Ohio led the nation in the number of sheep during the last half of the 1800s, there were several years in the 1860s and 1870s when sheep became a drug on the market. Aside from a few of the very best lambs, mutton did not sell at all and the price for fleece fell to an all-time low. Because the market for hides and tallow remained steady, John C. Baker, a man of keen perception in Mechanicsburg, Ohio, conceived the idea of buying sheep and slaughtering them for their hides and tallow. He opened an abattoir on the site of the present-day Goshen Park where he conducted a thriving business. The sheep carcasses he gave to local farmers who utilized them for fertilizer and the hides and tallow he sold at a profit.[9]

Mutton was one of the meats served in 1840 when William Henry Harrison, then a candidate for the President of the United States, campaigned in Champaign County. Following a two-hour speech, a meal was served in his honor on the lawn of one of the county's most prominent citizens. A local history says: ". . . twelve tables, each over three hundred feet long, had been erected and laden with provisions. Oxen and sheep had been barbecued."[10]

It was during the parade in honor of Harrison that the famous expression "OK" originated. Nearly every community in the county participated in the event. From Concord township came a wagon drawn by twenty-four horses. On the wagon was a board, one foot high by three and one half feet long, on which was painted "The People is Oll Korrect." In spite of the scoffing and accusations of ignorance by the opposing party, the term "OK" caught the public's fancy and today is heard and understood throughout much of the modern world.[11]

Directions for "Tanning Sheep-Skins, Applicable For Mittens, Door-Mats, Robes, Etc." was found in an 1867 book with the lengthy title, *Dr. Chase's Recipes; Information For Everybody: An Invaluable Collection Of About Eight Hundred Practical Recipes For Merchants, Grocers, Saloon-Keepers, Physicians, Druggists, Tanners, Shoe Makers, Harness Makers, Painters, Jewelers, Blacksmiths, Tinners, Gunsmiths, Farriers, Barbers, Bakers, Dyers, Renovators, Farmers, And*

Families Generally, To Which Have Been Added A Rational Treatment Of Pleurisy, Inflammation Of The Lungs And Other Inflammatory Diseases, And Also For General Female Debility And Irregularities All Arranged In Their Appropriate Department. It gave explicit instructions for cleaning and stretching a sheep's hide and for scraping off any remaining flesh. Then, it said, by rubbing the flesh-side with pumice or rotten stone, the skins would be "very white and beautiful, suitable for a foot-mat, also nice in a sleigh or wagon of a cold day. They also make good robes, in place of the buffalo, if colored, and sewed together. And lamb-skins, if the wool is trimmed off evenly to about one-half or three fourths an inch in length, make most beautiful and warm mittens for ladies, or gentlemen."[12]

S. P. Hildreth of Marietta, Ohio, related a unique cure using the carcass of a sheep. A man "was thought to be mortally injured from a falling tree, which caught him under the extreme branches, bruising his flesh all over as if whipped with a thousand rods. So many blows paralyzed the heart and rendered him as cold as a dead man. The doctor immediately ordered a large sheep to be killed and the skin stripped hastily off, wrapping the naked body of the man in the hot, moist covering of the animal. The effect was like a charm on the patient, removing all the bruises and the soreness in a few hours."[13]

Among the columns of advertising in the December 16, 1872, issue of "*The Monthly Visitor*," were interspersed bits of humor. One such item dealt with sheep: "A Scottish landlord was seated one day on the hillside of Bonally with a Scots shepherd, and observing the sheep reposing in what he thought the coldest situation, he observed to him; 'John, if I were a sheep, I would lie on the other side of the hill.' The shepherd answered: 'Ay, my lord; but if ye had been a sheep ye wad had mair sense.' "

Chapter 10: THERE WERE SHEEP

1. R. Douglas Hurt, *The Ohio Frontier* (Bloomington & Indianapolis, IN: Indiana University Press, 1996), p 225.
2. ——.
3. W. H. Beers, *History of Miami County* (Chicago: W. H. Beers & Co., 1880), p 405.
4. ——, *History of Madison County* (Chicago: W. H. Beers & Co., 1883), p 293.
5. *Annual Report of The Secretary of The State of Ohio, 1885* (Columbus, Ohio: R. Nevins Printer, 1886), pp 348-349.
6. R. Douglas Hurt, *The Ohio Frontier* (Bloomington & Indianapolis, IN: Indiana University Press, 1996), p 352.
7. *History of Miami County*, Miami County Ohio Sesquicentennial Committee (Columbus, Ohio: F. J. Heer Printing Co., 1953), pp 34-35.
8. S. P. Hildreth, *Biographical and Historical Memoirs of The Early Pioneer Settlers of Ohio With Narratives of Incidents And Occurrences in 1775* (Cincinnati, Ohio: Derby & Co., 1852), p 300.
9. Judge Evan P. Middleton, *History of Champaign County, Ohio* (Indianapolis, IN: Bowan & Co., 1917), p 1139.
10. ——, p 1132.
11. *Urbana and Champaign County*, Works Project Administration (Columbus, Ohio: Ohio State Archaeological & Historical Soc., 1942),p 41.
12. A. W. Chase, *Dr. Chase's Recipes; or, Information For Everybody: An Invaluable Collection of About Eight Hundred Practical Recipes, For Merchants, Grocers, Saloon-Keepers, Physicians, Druggists, Tanners, Shoe Makers, Harness Makers, Painters, Jewelers, Blacksmiths, Tinners, Gunsmiths, Farriers, Barbers, Bakers, Dyers, Renovaters, Farmers, and Families Generally, To Which Have Been Added A Rational Treatment of Pleurisy, Inflammatory Diseases, and also for General Female Debility and Irregularities: All Arrangedin their Appropriate Departments* (Detroit, MI: F. B. Dickerson Co., 1867), pp 219-200.
13. S. P. Hildreth, *Biographical and Historical Memoirs of The Early Pioneer Settlers of Ohio With Narratives of Incidents And Occurrences in 1775* (Cincinnati, Ohio: Derby & Co., 1852), p 13.

DOGS, CATS, RATS, AND "POLECATS"

Chapter 11

Dogs are thought to have been the first animals domesticated by humans in about 10,000 B. C. Artifacts from the earliest civilizations clearly show that man had formed a relationship with them. Almost every culture sooner or later appears to have established some kind of rapport with the four-legged canines. Explorers frequently took dogs with them. Hernando de Soto did. When he landed in the Americas in 1539, the terrified Indians never had seen creatures like the Spaniards' savage track and attack dogs—bloodhounds, wolfhounds, and mastiffs. European traders also were responsible in large measure for introducing dogs to different parts of the world. The first colonists on the shores of what was to become the United States brought their dogs. As people gradually pressed westward to settle in the newly opened Ohio country, their dogs ever were in tow.

Few persons headed for the Ohio frontier without their dogs. One historian wrote there were more dogs in early Ohio than livestock. In an 1885 annual report, it was estimated there were 168,398 dogs in the state.[1]

Dogs became an almost-integral part of the everyday life for people in the wilderness. The animals did service as protectors and as hunting companions. Some were enjoyed simply as pets. Dogs endured the same hardships as their masters, trailing along in their footsteps. The majority were large animals—bulldogs, Newfoundlands, mastiffs, and mixed-breeds. Some were part wolf. Their bodies often were scarred from numerous fights. It was not unusual to see dogs with ears missing, lost in battle with a bear or some other predator.

People set great store by their dogs' ability to detect danger. Pedro, an old dog who accompanied a group of emigrants from New Hampshire, kept a watchful eye over "his people." About 1810, in anticipation of an abundant corn crop, the men in the settlement decided to construct a grist mill on the Muskingum River. While one man took a turn at standing guard, the others worked. Old Pedro, always vigilant, instinctively positioned himself on an elevation nearby where at the least sign of a wild animal or an Indian he would begin to bark.[2]

While standing on top of a haystack, a pioneer in Gallia County was forking down feed for his cattle when he heard his dog suddenly begin to bark and show signs of concern. As he turned his head to look, he heard the unmistakable crack of a gun and a bullet whizzed past. Forty or fifty yards away, preparing to fire again, was an Indian. The man jumped from the haystack and sprinted toward the stockade, diving through the gate as another bullet thudded into the log wall close to his head.[3]

Late one afternoon in 1815 when he went into the woods to drive the cattle home, seven-year-old Robert Seeley of Lake County became hopelessly lost. The cows were determined to go in one direction and Robert was convinced "home" was in the opposite direction. The cows were right and Robert was wrong. For five days he wandered in the endless forest. Neighbors came from miles around to help search for the missing boy. Every now and again a dog that one of the families had brought with them would dart off into the woods and after a period of time, return. Each night, the dog disappeared. Finally, somebody noticed its curious behavior and decided to follow it. The dog led him straight to the child who told him that the dog had come to him at various times throughout the day and at night had stayed with him until daybreak.[4]

An 1888 newspaper carried the account of a dog who was responsible for saving a boy from drowning. Although youngsters in town repeatedly had been warned to stay off the thin ice on the local water works pond, young Joe Church could not resist the challenge. An engineer at the plant heard his St. Bernard frantically barking and when he investigated, found the boy in the water clinging to the broken ice. The men at the plant pulled him out, wrapped him in warm blankets and notified his parents.[5]

Until the breed became mongrelized, there was, in Jefferson County, a large, distinctive type of sheep dog. When the Marquis De Lafayette visited America in 1825 and toured parts of the country, he met with an Ohioan by the name of Bezaliel Wells. Among other things, the two men discussed the difficulties of raising sheep and Lafayette told Wells about the famous breed of dog in France that protected their flocks from wolves. He promised to send Wells a pair. "In due time they came and were quite prolific. They were a noble species, white with generally golden-hued spots; resembled the English mastiff. . . . One of them one day followed Alexander (Mr. Well's son), to market when a large, ferocious bulldog, encouraged by his master, attacked him. The butchers formed a ring around them expecting the bulldog to conquer. He had seized the shepherd-dog by the throat. The skin there was tough, and so loose that the other was enabled to twist his head around and grasp the bull's head, and soon the bones were heard to crack. The master of the bull then interfered. 'No,' said the others, 'we formed a ring to see fair play; you set him on and now we will see it out.' And they did. The shepherd-dog had got his spunk up, and they heard the crunching of the bones, and quickly the bulldog yielded up the ghost."[6]

There were occasions when dogs saved the lives of other dogs. One of these stories involved a party of hunters who took their pack of dogs into the woods with them as they scouted for game. A single dog was tied and left in camp. During the hunters absence, the dog in camp broke loose and with the tie dragging on the ground behind him, ran to join the others. Several miles from camp the strap snagged on the undergrowth and held him fast. Meanwhile, the hunting party was preparing to return to camp when one of the dogs seized a man by the sleeve and attempted to pull him in a different direction. Supposing the pack had brought down a deer or treed a bear, the men followed and found the trapped dog.[7]

An article about a Newfoundland and a mastiff appeared in a March, 1878, newspaper.. It told about two dogs that began snarling at each other over a bone they found lying on a bridge. The snarls escalated into a full-scale fight and while thus engaged, the dogs tumbled into the turbulent water below and were swept downriver. For several hundred feet the banks on both sides of the river rose sharply and there was no place for the dogs to climb out. The mastiff noticeably was beginning to tire when the Newfoundland managed to scramble ashore. Bystanders who were at the scene said the Newfoundland unhesitatingly sprang back again into the cold water. Swimming to the mastiff, he clamped his teeth on its collar and pulled him to safety.[8]

Although hunters often had dozens of hounds, each man had his preference. Simon Kenton, for example, had two favorite dogs he used for hunting. One day when a huge buffalo bull appeared on the path in front of them, both dogs rushed at it and seized it by its ears. The trio rolled down the icy hillside out onto the frozen river where the ice gave-way and the dogs, teeth still clutching the bull's ears, drowned. Simon lamented that "it was the greatest loss of his life—as it then seemed; for the dogs used to sleep each on one side of, and guard him of nights when [he] camped alone in the wilderness."⁹

A man in the southern part of Ohio said he liked dogs as well as the next man but he preferred they stay at home while he was hunting. They barked at the wrong time and scared away the game. They forever were being bitten by poisonous snakes or wounded by bears. When he was hunting, he wanted to hunt, not tend to injured dogs.¹⁰

George Davenport, an itinerant pewter-spoon molder who traveled from settlement to settlement plying his trade, had a dog, a big, ugly, but friendly dog that never strayed far from his side. One day in 1817, an Indian stole the dog and George followed, fiercely determined to reclaim his pet. He eventually caught-up with the Indian and found the dog. While searching for the Indian, George had had an opportunity to study the countryside and was so enthusiastic about the Darke County area he saw that he gave up his nomadic way of life and he and his faithful pet became permanent residents.¹¹

A man in Harrison County regaled anyone who would listen with tales of his Newfoundland-St. Bernard cross. The dog was an exceptionally bright animal and there was one example in particular the man liked to relate. Morning after morning, he said, it was obvious the dog had been outside the enclosed yard where he was kept. That in itself was not astonishing for the dog could, by pushing hard with his head, open the gate. But how it reentered the yard remained a mystery for the gate would open only one way. Finally, the man saw how it was done. The clever dog poked his nose through a knothole near the bottom of the gate and backed-up, pulling it open just far enough to squeeze through.¹²

Colonel Donn Piatt, who, in 1878, built Mac-O-Chee castle in Logan County, was fond of all animals and had numerous pets. He usually kept a minimum of half a dozen dogs and spent many pleasurable hours training them. Among the few who were allowed the freedom of the house was a diminutive black and tan dog he called "Frank." Although not overly affectionate, Frank was an intelligent dog and became a great favorite with the Piatts. Every morning, a bell signaled that breakfast was about to be served. The Colonel taught the dog to respond to the summons by accompanying each member of the household to the breakfast room, barking all the way, or as the family liked to say, "Frank barked them to breakfast." A man of countless interests, Piatt often became engrossed in the papers on his desk and would not immediately acknowledge Frank's summons. Frank then would dash out of the room and return with Tiny, another black and tan dog, and Nibbs, a Scotch terrier, and all three dogs would bark until the Colonel relented and went to breakfast.[13]

An 1878 newspaper printed an article that told about a lady who trained her pet poodle to carry a long stick in its mouth and on each end of the stick she would hang a lantern to light the path at night. During the day she frequently had the dog carry her sewing basket. One afternoon as she walked home from a neighbor's—the dog with the basket in its mouth,—a rabbit hopped across the road in front of them. Basket forgotten, the dog gave chase. The basket fell open and the contents scattered. Gathering up what she could find, the lady walked on home. Sometime later the dog appeared, holding a pair of shears in its mouth. It left again and reappeared with a spool of thread in its mouth. Again and again, it returned with something that had fallen out of the sewing basket. With the exception of a thimble, every item was retrieved.[14]

James Gallagher of Jefferson County told about a friend who owned a large dog that "one day strolled into the shop of one Peters, a butcher, and seizing a nice roast of beef, made off with it. Peters, on discovering whose dog it was, called upon" the owner who happened to be an attorney. The butcher "put the question to him: 'If a neighbor's dog enters my shop and steals a meat, is he not legally held in payment?' 'Certainly he is,' rejoined the attorney. 'Your dog,' continued Peters, 'has this very morning stolen seventy-five cents worth of meat from me and I have come for the money.' 'Not so fast, Mr. Peters,' replied the attorney. 'I don't give legal advice without compensation. As you are a neighbor, I won't be hard upon you. My charge to you in this case is $2.00. You must therefore pay me the difference of $1.25 and we will call it square.' "[15]

An Englishman who was visiting the United States had been invited to attend a dog show. Unable to attend, he expressed his regrets: "Your excellent American oysters, your roast beef, poultry and superior shad, have, I fear, caused a very provoking attack of gout, which will prevent me from visiting the Bench Show of Dogs, to open on Monday next. If the dogs to be exhibited prove to be no better than the dogs I have noticed along your streets, the exhibition will not prove very creditable. At least ninety in every one hundred dogs I have noticed in this city curl their tails to the left, an evidence of low breed and danger. Dogs that curl their tails to the right are never afflicted with hydrophobia; that fatal disease prevails only among dogs that curl their tails to the left. No gentleman in London or any city of the Continent will own a dog or allow a dog to follow him that curls its tail to the left."[16]

Unfortunately, there were dogs in the United States that contracted hydrophobia, regardless of the direction in which their tails curled. "Jennie Laycock, a twelve-year-old girl living at Mt. Gilead, Ohio, was bitten on the thumb by a dog and hydrophobia resulted. She is in critical condition," read a local paper.[17] Another publication printed: "Fourteen years ago, Daniel McKettrick, of Tiffin, was bitten by a dog. Last week, symptoms of hydrophobia appeared and soon the young man was stricken with spasms, frothed at the mouth and barked like a dog."[18]

July and August sometimes were known as Dog Days — Dog Days because it was the time of the year when the Dog Star, Sirius, one of the brightest stars in the sky, rose with the sun each day. There were people who erroneously believed that dogs were more likely to contract rabies during this period. Some further believed that swimming or using the water from lakes and streams where mad dogs drank would expose them to rabies.

There was an 18th century remedy for the bite of a mad dog which was reputed to be a "sure cure." "First, the patient must be blooded at the arm nine or ten ounces; then he must drink liverwort mixed in half a pint of warm cow's milk. After four doses he must go into the cold spring or river each morning for a month. Afterwards, he must go in three times a week for a fortnight longer."[19] Bleeding, i.e., the act of drawing blood from a patient, was an accepted medical practice.

Listed in the records at the Miami County courthouse are U.S. Army discharge papers for a "Private Trust." Private Trust actually was a dog that served from 1862 until the end of the fighting during the Civil War. He is said to have been the first dog ever mustered in and out of the Army. Both the dog and his master, Samuel Shannon, served under Ulysses S. Grant—Private Shannon in an artillery unit and Private Trust as a watch dog.[20]

Some people became so attached to their pets they memorialized them. One man commissioned a statue of his dog and arranged to have it placed on his burial plot in the cemetery to keep eternal watch over the family. Logan County's Colonel Donn Piatt interred his pet dog, Frank, in the Piatt mausoleum at the private graveyard on their property.

Piatt family mausoleum where Frank is buried.

photograph by author

Historical Collections of Ohio, by Henry Howe, includes accounts of Indians preparing pure white dogs with painstaking care and then killing them for sacrificial ceremonies. Chapters about both Seneca and Trumbull counties describe the ritual in considerable detail. The sacred rites when dogs were offered to the Great Spirit were solemn occasions followed with speech-making, feasting, and dancing.[21]

In Ashtabula County, a dog was hung. In the middle of the night on August 11th, 1812, a sentinel on duty rushed to the conclusion that incoming vessels from Lake Erie were invading British soldiers and dashed through the little settlement of Conneaut crying "Turn out! turn out! save your lives, the British and Indians are landing and will be on you in fifteen minutes!" The panic-stricken inhabitants hid in a dense thicket some distance away and because it was situated near the road, absolute silence was imperative. Every person complied; even the smallest babies were kept quiet. Dogs that had accompanied the villagers were silenced—all, that is, except for one young dog that excitedly continued barking. "Various means having in vain been employed to still him . . . it was unanimously resolved that that particular dog should die, and he was therefore sentenced to be hanged. With a sapling for a gallows, the young dog passed from the shores of time to yelp no more."[22]

Daniel Boone told about being held captive by Indians in 1778 and eating dogs to stay alive during a famine. When the dogs all had been consumed, they ate the bark on trees.[23] Hunters in the forest, white men and Indians alike, ate almost anything rather than starve to death. They devoured "the soles on their moccasins," and "even dog meat dripping with maggots," claimed one reference.[24]

Rufus Putman told about eating a dog when he was with a detachment of soldiers one cold winter. The "weather was excessively cold and the snow was five feet deep." There were thirty men in Putnam's detachment and their rations were gone. "By the sixth day they had no food except buds on beech trees and a few bush cranberries." The men were lame with frozen feet and so feeble, only a few of them were able to break a path in the snow. "The detachment happened to have a dog with them that was large and fat. . . . They killed him for supper and divided it into seven equal portions (one portion for every ten men). On the seventh day, some of the men breakfasted on one of the feet. A hind leg was cut off at the gambrel joint, which being pounded and roasted in the embers made a palatable meal." Putnam said the dog meat actually was very good and he "could eat that any time without disgust." He added that "it confirmed the experience of

Lewis' and Clark's men as they traveled over the Rocky Mountains who 'for weeks at a time lived on dog meat and preferred it to any other meat.' "[25]

There were 19th century factories that offered up to forty cents apiece for dead dogs. The skins were made into gloves; the hair was used in plaster; the bones were ground for clarifying sugar; and the fat was manufactured into oil. "Every part of the animal appears to be utilized, except its bark, and this, it seems to us, in the hands of a Yankee, might be saved and placed in the front yard to frighten off tramps and lightning rod agents," quipped a comedian.[26] S. B. Groves, a clothier in a small Ohio town, not to be outdone in the fashion world, advertised, "The best Dogskin Driving Gloves in the market can be found at S. B. Groves."[27]

In the event we should be tempted to ignore the fact there are individuals who harbor no special affection for dogs, the following bit of humor reflects that perception:

"Are you fond of dogs, Mr. Blevin?" asked a young woman as she caressed her pets.

Mr. Blevin was too polite to reply in the negative, so he murmured, "Oh, certainly."

"What kind do you like best?"

"I think," said Mr. Blevin reflectively, 'I like toy stuffed dogs better than most any sort.' "[28]

Immigration to Ohio with dogs and cats.
Historical Collections of Ohio, Henry Howe

****

Ancient Egyptians elevated cats to the status of sacred animals, identifying them with and worshiping them as gods. Temples were built to honor them; artists carved figures of them and made furniture and jewelry in the form of cats. When pet cats died, they often were buried in gold coffins. Their masters shaved off their eyebrows as a sign of mourning.

Although our Ohio ancestors had neither the time nor the inclination to venerate cats to the extent the Egyptians did, they did own cats and certainly brought them along as they ventured westward. Information about cats in early journals and diaries generally is in reference to their blood! Because cats' blood was a valued ingredient for medical remedies during the 17th and 18th centuries and the forepart of the 19th century, it was said that scarcely a barnyard cat could be seen that did not have a docked tail or a slit ear due to its frequent bleeding.

A typical entry in a 1786 diary read: "Mr. Davis came here and he has the shingles. We bled a cat and applied the blood that gave him relief."[29]

Even though barnyard cats far outnumbered the felines that were kept indoors, cats were popular as house pets. A Williams County, Ohio, history printed the following story: "One day I sat smoking and being busy in meditation I dropped off into a sort of doze. My cigar went out and I remained holding the stump between my lips. Seeing my somniferous condition, Tom gave a spring into my lap, crawled up to my face, and then turned partly round and with a poke of his paw knocked the stump out of my mouth on to the floor. Then he cuddled down into my lap and began purring. I never was more surprised. I felt almost like stopping smoking at the thought of a dumb animal like Tom teaching me such a lesson."[30]

Another cat warranted an item in an 1883 newspaper: "The window in a grocery store last week came down and caught a cat by the tail, while the groceryman was out; he heard a terrible noise in his store and hastened in. The window was raised and the cat left."[31]

****

Dwellings, barns and outbuildings frequently were infested with rats, those fast reproducing rodents that were imported to this country in the holds of the very sea-going vessels which transported our first colonists. Traps and poisons customarily were used in an effort to eliminate the noxious creatures.

An 1867 book offered several different poison recipes. One was titled, "Rat Destroyers" and gave directions for a thick paste using phosphorus and butter. It concluded by saying "Some will object to killing rats about the house; but I had rather **smell** their dead carcasses than **taste** their tail-prints, left on everything possible for them to get at, or suffer loss from their **tooth**-prints on **all** things possible for them to devour or destroy." The next recipe was called "Death For The Old Sly Rat" and told how to insert strychnine in small bits of meat. Another recipe, "Rats To Drive Away Alive," suggested dropping potash or cayenne pepper into their holes. "It will surely send them off at a sneezing pace." The last recipe, "Rat Poison—From Sir Humphrey Davy," was said to be "tasteless, odorless and infallible. It will prove just the thing for rat-killing, as they can be gathered up and carried away, thus avoiding the stench arising from their dead carcasses."[32]

The November 27, 1867 issue of the *Urbana Union* printed: "What It Costs To Shoot A Rat In Urbana—Our old friend and well-known citizen of Urbana, Mr. Harry Marsh, on Monday last, fired a gun at a rat in the cellar of his premises, No. 26 North Main St. The consequence was that Marsh was taken in charge by the Marsh-al and brought before Mayor Long who, by virtue of an ordinance of fifteen years standing, fined Mr. M. $2 and costs, amounting altogether to $4.20. It seems that the ordinance aforesaid forbids the firing of any kind of arms, within the corporation limits, whether or not the firing be on the premises of the party offending. We suppose very few persons were aware of it and we think the city fathers should pay Mr. Marsh a liberal sum for being the means of having it published."

Hotels were not exempt from the problem of rats. The Exchange Hotel, identifiable today as the Hotel Sowles, has been listed as one of the oldest hotels in the State of Ohio.

Rats were a nearly constant source of irritation. An 1884 newspaper even carried an article about a hotel employee catching rats. "A dining room boy being troubled with rats and not caring to use poison, concluded to catch the rats himself. He secured an old coat and threw it over the hole where they got into the room - first putting crumbs of bread and cheese on the floor - and then kneels down in front of the holes and awaits his ratship's coming, which was not long and when the rat gets fairly into the room, the boy rolls him up in the coat. He had just caught one when our reporter was there."[33]

****

Skunks, more commonly called "polecats" by our forefathers, are black and white carnivores that belong, not to the cat family as one might be tempted to suspect from its name, but to the weasel family. David Zeisberger, the Moravian missionary who made frequent references to Ohio wildlife in his book about Delaware Indians in Ohio, said a skunk had "a gentle and mild countenance" but "it goes out of the way for no one and whoever approaches too near is ill rewarded for his curiosity."[34]

Skunks do give warning, however, before activating their defense mechanism. When the creatures feel threatened or alarmed they raise their tail and scratch or pound on the ground. If the warning is not heeded, they hiss. If the "threat" still does not retreat, it becomes the recipient of an astonishingly accurate spray—up to ten feet—of "skunk juice." The spray, a vile, oily, yellow substance, is directed at the face where the fluid stings the eyes, sometimes resulting in temporary blindness.

A twelve-year-old-boy while checking his trap-line along a creek, saw a skunk just ahead of him about to enter its burrow. He ran forward and with the forked stick he carried, pinned the animal's tail to the ground. Then he realized his predicament. Only the tail was visible; the rest of its body was inside the burrow. One of his hands was holding the stick firmly in place and he knew he didn't dare let loose. Because he wasn't quite certain what to do, he finally sent the dog home, knowing his father would understand he needed help of some kind. His father did immediately grasp the situation and, grabbing his gun, followed the dog. Father and son worked together, pulled out the skunk and killed it.[35]

Ansel Blossom settled in Van Wert County and soon thereafter became a Justice of the Peace. One day, shortly before officiating at a wedding ceremony, he went into his milkhouse and saw a polecat drinking milk from a pan. He dashed to the house for a fire shovel and quietly returning to the milkhouse, "brought the shovel down upon his neck, cramming his head into the milk, intending to drown him; but the animal gave him such a sprinkling as to render him blind for a time, and to perfume his clothes . . . ." Mrs. Blossom buried her husband's clothing deep in the ground. The wedding was postponed.[36]

While a polecat's stench is unbelievably offensive, its flesh is not. Indians found that the meat was quite savory and had no unpleasant odor.

Many of the settlers hung a bag of skunk oil and camphor around their neck to ward off boils and sore throats. People sometimes made pouches by stripping off the polecat's skin and cutting a small slit at the neck for an opening. Trappers occasionally marketed skunk fur as "Alaskan sable" and there always were furriers who were not adverse to using the pelts.

Normally, polecats tended to avoid dense forests, seeking instead, grasslands and cultivated regions near human habitation—human habitation that far-too frequently afforded the unwelcome mammals the opportunity to slip into chicken houses and eat the eggs. Skunks, on the other hand, moving in their slow, clumsy-like gallop, actually were of some slight benefit to the settlers for included in their varied diet were rodents, bees, wasps, and snakes. As part of a novel 1872 advertisement for onions, an inventive store keeper used the following composition.

"The Pole Kat

My friend, did you ever examine the fragrant pole kat closely? Their habits are phew but unique—
They are called pole kats bekause it iz not convenient tew kill them with a klub, but with a pole, and the longer the pole the more convenient.
When a pole kat iz suddenly walloped with a long pole, the fust thing that he, she, or it, duz, iz tew enbalm the air for menny miles in diameter with an akromonious ollfaktory refreshment, which permeates the ethereal fluid, with an entirely original smell.
One pole kat in a township iz enuff, espeshily if the wind changes once in a while.

Billings"[37]

Chapter 11: Dogs, Cats, Rats, and "Polecats"

1. *Annual Report of The Secretary of State of Ohio, 1885* (Columbus, Ohio: R. Nevins Printer, 1886), pp348-349.
2. S. P. Hildreth, *Biographical and Historical Memoirs of the Early Pioneer Settlers of Ohio With Narratives of Incidents and Occurrences in 1775* (Cincinnati, Ohio: Derby & Co., 1852), pp 453-454.
3. ——, p 375.
4. *Ohio Historical Review*, vol. 9, no. 45 (Columbus, Ohio: pub./ed. Ralph S. Estes, n. d.), p 10.
5. *Urbana (Ohio) Citizen and Gazette*, December 27, 1888.
6. Henry Howe, *Historical Collections of Ohio*, vol. 1 (Norwalk, Ohio: Laning Printing Co., 1896), pp 974-975.
7. *Urbana Citizen and Gazette*, November 14, 1878.
8. *Urbana (Ohio) Citizen and Gazette,* March 14, 1878.
9. Edna Kenton, *Simon Kenton, His Life and Period 1755-1836*, (New York, N.Y.: Random House from the Doubleday edition, 1930), p 72.
10. William Steele, *The Old Wilderness Road, An American Journey* (New York, N.Y.: Harcourt, Brace & World, 1968), p 77.
11. W. H. Beers, *History of Darke County* (Chicago: W. H. Beers & Co., 1880), p 415.
12. Henry Howe, *Historical Collections of Ohio*, vol. 1 (Norwalk, Ohio: Laning Printing Co., 1896), p 891.
13. C. G. Miller, *Donn Piatt: His Work and His Ways* (Cincinnati, Ohio: Robt. Clarke & Co., 1893), pp 297, 298.
14. *Urbana (Ohio) Citizen and Gazette*, April 4, 1878.
15. Henry Howe, *Historical Collections of Ohio*, vol. 1 (Norwalk, Ohio: Laning Printing Co., 1896), pp 971-972.
16. *Urbana (Ohio) Citizen and Gazette*, May 29, 1879.
17. ——, August 18, 1884.
18. *St. Paris (Ohio) New Era*, April 14, 1879.
19. Julianne Belote, *Compleat American Housewife 1776* (Conrad, CA: Nitty Gritty Productions, 1974), p 135.
20. Damaine Vonada, *Amazing Ohio* (Wilmington, Ohio: Orange Frazer Press, 1989), p 41.
21. Henry Howe, *Historical Collections of Ohio*, vol. 2 (Krehbiel & Co., 1904), pp 575-576, 664.
22. ——, vol. 1, pp 278-279.

23. Orton G. Rust, *History of West Central Ohio* (Indianapolis, IN: Historical Pub. Co., 1934), p 177.
24. ——, pp 194, 195.
25. S. P. Hildreth, *Biographical and Historical Memoirs* . . . (Cincinnati, Ohio: Derby & Co., 1852), pp 13-18.
26. *(St. Paris, Ohio) New Era*, July 26, 1877.
27. ——, April 19, 1877.
28. *(Urbana, Ohio) Monthly Visitor*, vol. 17, no. 1, March, 1889.
29. Rebecca Rupp, "Animal Friends," *Early American Homes* (Feb., 1997), p 49.
30. Henry Howe, *Historical Collections of Ohio*, vol. 2 (Cincinnati, Ohio: Krehbiel & Co., 1904), pp 855-856.
31. *St. Paris (Ohio) New Era*, Nov. 16, 1883.
32. A. W. Chase, *Dr. Chase's Recipes*. . . . (Detroit, MI: F.B. Dickerson Co., 1867), pp 320, 321.
33. *Urbana (Ohio) Daily Citizen*, Sept. 11, 1884.
34. Rev. David Zeisberger, *David Zeisberger's History of The Northern American Indians*, ed. Archer Hubert & William Schwarz (1910; reprint, Lewisburg, PA: Wennawoods Pub., 1999), p 62.
35. Author's conversation with her father, Philip E. Stickley re family history.
36. Henry Howe, *Historical Collections of Ohio*, vol. 2 (Krehbiel & Co., 1904), p 725.
37. *Urbana (Ohio) Citizen and Gazette*, Urbana, Ohio, March 28, 1872.

ET CETERA CRITTERS
Four-Legged Ones

Chapter 12

While skunks are considered fur-bearing animals, they were not the species which initially attracted fur-trappers to the Northwest Territory. The entire region was an incredible hunting ground for badgers, weasels, minks, coyotes, and both gray and red foxes. The creatures the American and European markets most desired, however, were beavers and this was the primary reason trappers came to Ohio.

In search of the beaver, men leading strings of packhorses followed game trails into the unexplored wilderness. Some of the men came in long, hollowed-out log canoes, paddling along the water-ways. In defiance of the sometimes friendly and sometimes hostile relations with Indians, traders soon began establishing trading posts in Ohio, first on the shores of Lake Erie, particularly in the Maumee River Valley and around Sandusky Bay.

One historian has said the beaver was the single most important animal in the development of commerce in the New World.[1] Beaver coats were admired for their warmth and for their sleek, elegant appearance. They became a status symbol for the aristocracy and members of high society.

Beavers were inoffensive and defenseless. They had poor eyesight and a comical way of walking. They were buck-toothed and paddle-tailed. "Their trowel-shaped tails were so heavy that they dragged the ground, like a board dragged along by one end," wrote James Swisher.[2]

Notwithstanding some of their physical attributes, beavers had beautiful, soft, dark brown coats and it was these the purchasing public wanted. Often referred to as "hairy bank notes,"[3] the price of their pelts varied widely but some have estimated that they averaged from between $4.00 to $9.00 per pound.

Not only did beaver skins sell well but so, too, did their musk glands which were thought to have medicinal value. Although there was only slight market for beaver tails, many of the hunters and trappers ate them, considering them a delicacy. Beaver tails, which were nothing more than greasy flaps of gristle and fat, were wrapped in wet oak leaves and left to bake all night in a bed of hot coals in the fire.

In the mid 1700s, James Smith, a young man who had been captured and adopted into a tribe of Delaware Indians, told about a belief held by a tribe in the northeastern part of Ohio. Some of the members in the tribe were of the opinion geese could turn into beavers. One of the elders said that the pond near where they were camping "had always been a plentiful place of beaver. Though he said he knew them to be frequently all killed, yet the next winter they would be as plenty as ever. And as the beaver was an animal that did not travel by land, and there being no water communication, to, or from this pond—how could such a number of beavers get there year after year? But as this pond was also a considerable place for geese, when they came in the fall from the north, and alighted in this pond, they turned into beavers, all but the feet, which remained nearly the same."[4]

Smith wrote in his memoirs about having a pet beaver while he lived with the Delawares at the beaver pond. When his Indian father asked him if he knew why beavers built dams, he replied that he supposed beavers built them in order to create ponds so they would have ready access to a supply of fish to eat. His Indian friends laughed uproariously at the idea of beavers eating fish. Smith was perplexed. He offered fish to his pet beaver and found that it refused not only fish but also all other kinds of flesh. Still not persuaded that beavers did not eat fish, he killed one of the animals and examined the contents of its stomach. There was no sign of fish, only the remains of the bark of trees, roots, and certain other vegetation.[5]

****

Raccoons, those greedy, mischievous animals with fox-like faces and a black patch around each eye which give them the appearance of a bandit, were heartily disliked by the early pioneers. They could ravage a corn crop in short order. Although Indians willingly ate the meat, few white people truly relished it. Simon Kenton told of having to live on raccoon meat for two weeks as he made his escape from the British garrison at Detroit and traveled southward toward his home. "[We] often had as high as three of a night.[6]

Reverend Finley left a description of a meal he had shared with Indian hosts in their camp that consisted of nothing except raccoon meat. "Soon we had placed before us a kettle filled with fat raccoons, boiled whole, after the Indian style, and a pan of good sugar molasses. These we each carved for himself, with a large butcher knife. I took the hind quarter of a raccoon, and holding it by the foot, dipped the other end in the molasses and ate it off with my teeth. Thus I continued dipping and eating till I had pretty well finished the fourth part of a large coon."[7]

Jonathan Alder wrote about eating coon while he lived with the Indians. Game was nearly non-existent one piercingly cold winter. "Even the raccoons were lean," he said, "which made very poor eating." He explained that "we did not even skin the coons for fear of losing some part of them that could be eaten. We generally threw them into the fire to singe off the hair, and then ate them hide and all! Some of them were so poor that I have frequently seen our dogs refuse to eat them! Yet, for months we were compelled to eat them or starve!"[8]

Captain Robert Benham was one of two survivors of a well-laid Indian ambush in the autumn of 1779. Shot through both hips and unable to walk, he dragged himself to a fallen tree and hid as the Indians pursued others in his party. The following day when the Indians returned to strip the dead, he remained undetected. Sometime during the second day he killed a raccoon as it descended a tree near him and while he was trying to devise some way of reaching it, he heard someone call to him. Suspecting it was an Indian, he reloaded his gun and waited. Presently, he heard the same voice again, only much closer. Still, Benham made no reply but cocked his gun and was prepared to fire as soon as the Indian showed himself. The voice called a third time, impatiently and obviously in some distress. Feeling more confident that the voice was of another person in his group, the Captain responded. The man who appeared was, indeed, one of his men and he, also, had been wounded—both arms were broken.

There is a long story in Volume Two of Henry Howe's *Historical Collections of Ohio*, about the men. One could not walk and the other had no use of his arms. Captain Benham killed game, most frequently raccoons, and the other man used his feet to kick it to the place where Benham sat. The man with the broken arms obtained water by wading into the creek holding his hat in his teeth. He dunked his head under the surface, filled the hat with water and carried it back to the Captain. In November, they were rescued and taken to Louisville.[9]

Raccoons, celebrated tree climbers, often live in the hollows of trees. When Smith Hill and several of his nephews tracked a coon to a hollow in a tree in Van Wert County, they began to chop it down. Shortly, their dogs ran to another tree and commenced barking at something in the treetop. Hill leaned his gun against the trunk of the tree and, carrying the axe, went to investigate. Suddenly, a "very large bear came at him." As he backed-up to try to escape, his heel caught in the brush and he fell. By the time he regained his feet the bear was "almost onto him and he was compelled to fight. He sank the bit of the axe into its head and killed it. It was very large and very fat."[10]

An early Ohio farmer related that, raccoons "did a great deal of mischief to the corn in the summer, eating it, and breaking it down. We often hunted them of nights. They would come into the corn fields soon after dark. Then we would send in our trained dogs. The raccoon would seek refuge on the largest tree it could find. A trained raccoon dog has a peculiar kind of bark when he trees the animal, which the hunter at once recognizes. If the tree was not too large, we at once cut it down. The dogs would be ready near where it would fall, and rarely missed catching and killing the raccoon at once. If the tree was very large, we would build a fire, roast the green corn, tell stories, and thus amuse ourselves until daylight, when we would shoot the raccoon, and thus save the labor of cutting it down."[11]

A group of hunters once spied a coon in a tree and persuaded one of the members to climb the tree. Telling him to shake the limb so the coon would tumble to the ground for the dogs to catch, they stood below and watched. "Accordingly, he ascended, and stealing softly from branch to branch, in search of the coon, he finally espied him snugly ensconced on one of the topmost branches. . . . Proceeding cautiously, he reached the limb below that on which was the coon. Raising himself up for the purpose of reaching the limb which he intended to shake, the one on which he stood was heard to crack and began to give way. He was now thirty feet from the ground. Aware of his perilous condition, he cried out to his companions below, 'I'm falling.' Seeing his danger, and that nothing scarcely less than a miracle could save him from death, they besought him to pray. 'Pray,' said he; 'I haven't time; I can't pray.' 'But you must pray. If you fall, you will be killed.' He then commenced repeating the only prayer he knew: 'Now I lay me down to sleep;' but he could proceed no further, as the cracking of the limb indicated its speedy severance from the trunk, and he cried out at the top of his voice, 'Hold the dogs; I'm coming.' And sure enough, down he came with a crash; and the dogs, thinking it to be the coon, were with difficulty restrained from attacking the coon-hunter, who was considerably stunned by the fall."[12]

It was not uncharacteristic for coons to escape their pursuers by climbing a large tree and then crossing on its branches to the overlapping branches of an adjacent tree, leaving the dogs excitedly barking at the base of the wrong tree. The phrase, "barking up the wrong tree," is thought to have originated in this way.

Many magnificent trees in the unbroken Ohio wilderness were destroyed for the sole purpose of killing a raccoon. One man wrote that he often had seen these immense trees set on fire merely to dislodge a single scrawny raccoon!

Tall tales were a frequent source of entertainment along the frontier and the tales became ever more outrageous with each retelling. A man by the name of Blevins told a story about a tree and a host of raccoons. He said "he remembered once finding a hollow tree that acted in a mighty strange way. It grew big and then shrank, puffed out big, then grew small. He figured it might be the devil inside, so he cut the tree down and out came two dozen or so raccoons. He was so surprised that he let them all get away without killing a one. They must have been packed so tightly in the hollow tree that when they breathed in, they swelled. When they let their breath out, the tree would shrink."[13]

Raccoons have easily recognizable cries so when Thaddeus Gilliland heard the familiar sound one night, he picked up his gun, went outside, killed the furry animal and tossed the body into the smoke house. Scarcely had he stepped inside again when he heard another raccoon squall. He killed it, tossed it into the smoke house and started for the cabin. Nine times the same thing occurred. As a tenth raccoon began creating a fuss, Gilliland decided he was too tired to bother with more and went to bed. Next morning, much to his surprise, he found that during the night his dog had killed the tenth one and after depositing it against the door, lay there waiting for him.[14]

From 1838 until 1844, a raccoon was the emblem for the Whig party in the United States. Thus, in 1840 when William Henry Harrison campaigned for the Presidency, a raccoon was the established symbol. As Harrison traveled through the nation, he was presented as "the log cabin, hard cider" candidate, all designed to demonstrate his rugged Ohio background, thereby implying the depth of his understanding of frontier concerns. Barrels of cider and small log cabins on horse drawn wagons were featured in long parades. Nailed to the exterior walls of the cabins were rows of stretched coonskins. Live coons, which as a rule were trained with little difficulty and made fine pets, were secured to the cabins' ridgepoles where they scampered up and down the roofs, amusing the crowds with their antics. Men wearing coonskin caps and leading tame raccoons on leashes marched in organized groups.[15]

Coonskin caps were popular wearing apparel with backwoodsmen. Sometimes the head gear was decorated with the coon's tail, sometimes with a fox's or a squirrel's tail hanging down the back.

Indians made coonskin containers and used them to carry liquids. Simon Kenton, the famous frontiersman, said that he often purchased bear's oil in coonskin bags from the Indians.[16]

Because fur traders were willing to pay from 25¢ to 50¢ per raccoon skin and because taxes could be paid in furs, which included raccoon fur, large numbers of the animals were killed. It was said that in the 1800s alone, over one million were killed every year.

Since opossum skins virtually were worthless, unethical hunters would cut the tails off raccoons and sew them onto opossum hides, rolling them into bundles with the coon tails exposed. Many a gullible county treasurer paid for opossum skins thinking they were raccoon.[17]

The *History of Hardin County, Ohio*, said that John Wilson of Liberty township dealt in furs, primarily coonskins. Few knew his given name was John. Everybody called him "Coonskin Wilson."[18]

On February 2, 1804, the State of Ohio granted a charter for a library to a group of civic-minded citizens in Athens County who then had the formidable task of raising money to purchase books. There was considerable interest but subscriptions were small for money was scarce. "So scarce was money," said Judge A. G. Brown, "that I can hardly remember ever seeing a piece of coin till I was a well grown boy. It was with great difficulty we obtained enough to pay our taxes and buy a little tea for mother."[19]

It was suggested that people who wished to participate in funding the library might hunt and trap and donate whatever amount they could afford from the sale of the pelts. This was done and eventually one hundred dollars was collected, almost entirely from the sale of raccoon skins. One man said, "All my accumulated wealth, ten coonskins, went into the fund."[20]

Two representatives of the proposed library traveled to Boston where they deliberated at length, finally selecting fifty-one books they felt were appropriate. They wrapped each book carefully and by way of packhorses transported them to Ames, the site chosen for the library in Athens County. In view of the fact so much of the fund had been made possible through the sale of coonskins, the library almost immediately became famous as the "Coonskin Library." "There never was a library better read," reported Henry Howe, author of *Historical Collections of Ohio.*[21]

****

For farmers, squirrels were a serious pest in the early-to-mid 1800s. The destructive rodents could, and oftentimes did, totally destroy crops of corn. Pioneers organized sporting events for the purpose of eliminating squirrels. A community a short distance south of Troy, Ohio, planned a squirrel hunt and offered prizes. The individual who killed the greatest number was offered a premium of one hundred and fifty bushels of corn; second place was one hundred bushels; and third place was fifty bushels. Only expert marksmen were eligible to participate. Each contestant was permitted to carry two rifles and one person could accompany him to load the gun; one person could scalp the squirrels; and another could help drive the furry little varmints toward him. At the end of six days the scalps were counted. The first prize was won by a man who killed 1,700 squirrels. The second place winner killed 1,300 squirrels.[22]

Elijah Kiser was a renowned squirrel hunter and once rated special attention in the newspaper: "There lives in the locality west of Northville, Elijah Kiser, a mighty hunter who can knock the eye out of a squirrel at eighty yards."[23]

People who could afford it, kept two rifles, a heavy caliber for large game and a smaller caliber for squirrels and small game. Many of the old-timers adapted their well-worn six foot long rifles for squirrel hunting. Almost every man and boy was an accomplished shot and oftentimes women and girls also became excellent marksmen.

It required unrelenting surveillance to keep the scavengers away from crops. Parents delegated their children to stand guard and walk around the fields making loud noises. One father carved whistles from a pawpaw bush for his youngsters to blow as they patrolled. It was a dawn-to-dusk duty.

One ten-year-old boy's job every day during the planting season was to march around the cornfield beating on an old tin pan to keep squirrels away. [24] Children old enough to handle a gun were given free rein to practice their shooting skills by aiming at squirrels.

Efforts at guarding the fields had to be resumed in the fall when the corn ripened. Again, children were called upon to help. Saving one's crops was of the utmost importance to Ohioans for it could make the difference between having something to eat or starving.

Every few years there were migrations of squirrels, thousands at a time, searching for food. In some places the destruction was likened to an invasion by an army that laid waste every foot of territory through which they swept. "The woods were fairly alive with them. Thousands must have been under our view without turning our head."[25]

The squirrels "made their journey in the fall of the year, about the time the corn began to ripen. They appeared in such vast numbers, as apparently to cover the earth for miles, and if not well guarded, they would clear the corn fields as they went along. They would pass over houses and swim lakes, ponds and water courses."[26] When they reached the Ohio River, they "would plunge and attempt to swim over; here an immense number would lose their lives by drowning in the river and those that got over alive would crawl up on the bank and after resting a short time, would resume the journey southward."[27]

In 1822, there was a report that the squirrels were so ravenous "they forgot all fear and if a person had an ear of corn in his pocket, they would swarm upon him and cut the very clothing with their teeth to get at the corn they were so famished."[28]

Mr. Longfellow, who settled near Eris, Ohio, in order to secure his crop one season, "hauled his entire crop to his house and stacked it around the yard. Coming out of his house one morning a drove of, perhaps, a hundred or more were at work at his corn. He called his dog and chased them away, sixteen beating a retreat up the well-pole."[29] Interestingly enough, although the gentleman's name was Longfellow, he stood only four and one half feet tall.

In Van Wert County, a farmer, despairing of ever ridding his field of the squirrels, supplied the ammunition for a group of four men who in one day killed two hundred and forty squirrels. It was a fruitless task, nonetheless, for in a very short period the squirrels "seemed to be as plentiful as ever."[30]

The disheartening loss of crops prompted concerned officials in Adams County to pass legislation requiring every male over the age of twenty-one to produce one hundred squirrel scalps or pay three dollars in CASH![31] In 1808, Township trustees in Salem township, Champaign County, charged all taxable citizens ten squirrel scalps or one scalp for each and every twelve and one half cent tax due.[32] When Brown County property was appraised for taxes in 1809, each taxpayer was informed that in addition to paying his taxes, he also would be assessed a certain number of squirrel scalps. Failure to comply cost three cents for every scalp that was in default. Certificates were issued allowing credit of two cents per scalp for excess scalps. These certificates were accepted by merchants in exchange for goods.[33]

The last known migration of squirrels to pass through several of the western counties occurred in 1836. They came from the northwest, crossing in a diagonal direction. "The 'stampede' continued about a week, but was at its height not longer than twenty-four hours."[34] During the Civil War, when Cincinnati was threatened by an attack from the approaching Confederate army, thousands of emergency volunteers from all over Ohio rushed to the town's defense. Many in the hastily summoned guard were expert squirrel hunters and the militia forever after was christened the "Squirrel Hunters."[35] It was noted that a large percentage of the men wore caps from which dangled squirrel tails. Others dressed in buckskin shirts decorated with squirrel fur.

****

In 1794, soon after arriving in Ohio , John Johnson constructed a primitive grist mill. "After the mill was ready for work, it was not used for some days, but that at length a neighbor brought some corn to be ground, which was put into the hopper and the mill started. The corn disappeared but the meal did not appear. This puzzled Mr. Johnson very much and, at length, he took the mill apart to see what the difficulty could be and found that some ground-squirrels had built a nest in the mill and the corn was ground so slowly that they would eat the meal as fast as it was ground."[36]

****

Also known as "woodchucks," groundhogs, with their whiskered, broad, flat heads, were nearly as unpopular with settlers as raccoons and squirrels. Livestock sometimes stepped into groundhog holes and broke legs, which meant the animals had to be destroyed.

Groundhogs had no economical value. Few people considered their flesh edible and their hair was too coarse to make acceptable fur. Trappers and hunters aspired for fur-bearing mammals such as beavers, otters, muskrats and raccoons—those whose pelts could be sold. Primarily because their hide did have no monetary worth, farmers kept and tanned groundhog hides for themselves to make straps and tugs, bags and pouches, i.e., items for their own utilitarian everyday use.

Indian Joe, a crippled Shawnee, was a master tanner. When the rest of his tribe moved onto a reservation, Indian Joe refused to leave his home at Old Chillicothe (now known as Oldtown, Ohio) and for the remainder of his life made his home with Aaron and Hannah Paxson in what today is Greene County. This expert tanner, "especially of groundhog skins, some with the hair or fur on, and some without it, called the fine, tough product, 'whang leather.' . . . Old Joe's primitive method produced a perfect tan of all native wild animal hides and he was noted for the durable leather he made. . . . He always used the brains of the slain animal in his process. When the neighbors brought him a groundhog skin to tan, he always had them bring, at least, the head of the animal, if not the whole carcass." Then "he would sit for hours at a time and manipulate this skin and brains over and over until the entire brain substance was completely absorbed into the skin, after which the leather remained pliable, and was not, apparently, affected by time or weather."[37]

Indian Joe once told a Paxson grandson "that the sun came up out of a groundhog hole up at Hickory Point, a small grove on the eastern side of the grandparents farm, and that it set behind a big stone on Steel's Hill, where he said the sun went behind the rock to the westward. He showed the boy a groundhog hole up at those woods, where it had been worn smooth, and he said that it was the sun coming out that made it smooth."[38]

For those persons who cared to know, an 1880, Noble County newspaper printed the directions for tanning the skins.

"To Tan the Hide of a Ground-Hog

After taking the hide from the animal remove as much of the flesh as possible. Then take fresh wood ashes and put into a vessel, and put in enough water to make a thick mortar. Then put in the hide and let remain until the hair gets loose. Now remove the hair and place the hide in a vessel of soft water to soak, and let it so remain about three days. Then take it out and wash it, and stretch it as it dries. Particular care must be taken not to allow it to stay too long in ashes, as the ashes will eat the hide up. The hair generally comes loose in about three days.

Tim"[39]

Much as the bushy tailed groundhogs were maligned, there were relatively few adults or children who were not familiar with and did not enjoy the simplistic jingle about a groundhog.

How much wood would a woodchuck chuck,
If a woodchuck could chuck wood?
A woodchuck would chuck as much wood
As a woodchuck could chuck wood.

****

Only a few short accounts pertaining to porcupines could be found for this collection of pioneer and critter stories. Henry Howe mentioned the first incident which occurred in Putnam County. He wrote about a man by the name of Sebastian Sroufe who thought he had discovered a strange, new animal. Stroufe related to Howe: "I sprang from my horse and killed it with a club, it showing no fight. I then tied it on my horse, back of the saddle." That proved to be an ill-advised move for a porcupine's back, sides and tail are covered with quills, quills covered with many, tiny, fishhook-like barbs. Almost immediately Stroufe's horse became unmanageable, "dancing up and down, especially the back part of him; then trotting off." Eventually, Stroufe caught his mount and "saw what was the matter. A quantity of pin-quills were sticking in his back, gathered from the animal. Every motion of his body drove them farther in." The porcupine was removed but the barbs had broken off and remained embedded, causing the horse considerable suffering.[40]

It was reported that dogs who attacked porcupines usually died unless all the quills were quickly and completely removed. Although removing the deeply anchored, backward facing barbs was vital, it was not easily accomplished.

Henry Howe, while continuing his history of Putnam County, told how Native American women utilized porcupine quills for decorative purposes.[41] Reverend Zeisberger, in his *History of The Northern American Indians,* observed that Indian women used the hollow, sharp spines to adorn moccasins and pouches. To make them colorful and more attractive, they often dyed them a bright red.[42]

The easy-going porcupine, with its thirty thousand quills, devoted most of its time to eating and sleeping. It was not a carnivorous animal but, instead, ate the bark, buds, and twigs of certain kinds of trees. Rarely did the early pioneers eat porcupine flesh. America Indians frequently did.

According to a history of Washington township in Montgomery County, the last porcupine, or "American hedgehog," as it sometimes was called, was killed one Sunday morning in the summer of 1830, while crossing the street in Centerville.[43]

****

The Reverend James B. Finley spoke about opossums in his autobiography, calling them "ugly and deceitful. If you strike him, he will roll over, and appear as if dead, and as soon as you leave him, he starts up and hastens to his den. His tail is entirely bare and serves many good purposes to the animal. He is not easily shaken off a tree like the coon but clings to it with the greatest tenacity, winding his tail around the limb and defying all efforts to shake him down. A hard-shell Baptist preacher once introduced this animal into his discourse to illustrate the doctrine of final perseverance. . . .The flesh of this animal is like that of the young pig; his oil is abundant and answers well to burn in lamps or grease harness. The flesh of the opossum and new corn mush was considered a most delicate dish among backwoods families. Their skins, when dressed, are as white as the skin of the chamois and make fine gloves for backwoods ladies."[44]

Chapter 12: ET CETERA CRITTERS: Four-Legged Ones

1. Robert McCracken Peck, *Land of The Eagle* (New York, N.Y.: Summit Books, 1990), p 62.
2. James Swisher, *How I Know* (Cincinnati, Ohio: Press of Jones Bros., 1880), p 227.
3. Karen Terry, "A Passion For Beaver Pelts," *Cobblestone* (Dec., 1991): p 32.
4. Col. James, Smith, *An Account of The Remarkable Occurrences in The Life and Travels of Col. James Smith, During His Captivity With The Indians in The Years 1755, '56, ' 57, ' 58, and '59* (Cincinnati, Ohio: Clarke & Co., 1870), pp 57, 58.
5. ——, pp 58, 59
6. Edna Kenton, *Simon Kenton, His Life and Period 1755 -1836* (New York, N.Y.: Random House from the Doubleday edition, 1930), p 140.
7. James B. Finley, *Life Among The Indians* (Cincinnati, Ohio: Hitchcock & Walden, 1857) p 295.
8. Johnda Davis, *Journal of Jonathan Alder* (Columbus, Ohio: Ohio State Historical Society, 1988), pp 60-61.
9. Henry Howe, *Historical Collections of Ohio*, vol. 2 (Cincinnati, Ohio: Krehbiel & Co., 1904), pp 741-42.
10. Thaddeus Gilliland, *History of Van Wert County* (Chicago: Richland & Arnold, 1906),pp 149-150.
11. *Ohio Historical Review* (Columbus, Ohio: pub./ed. Ralph S. Estes, 1990), p 62.
12. James B. Finley, *Autobiography of Rev. James B. Finley or Pioneer Life in the West* (Cincinnati, Ohio: Cranston & Curts, 1853), pp 94-95.
13. William Steele, *The Old Wilderness Road, An American Journey* (Orlando, FL: Harcourt, 1968), p 98.
14. Thaddeus Gilliland, *History of Van Wert County* Chicago: Richland & Arnold, 1906), p 139.
15. Henry Howe, *Historical Collections of Ohio*, vol. 1 (Norwalk, Ohio: Laning Printing Co., 1896), p 377.
16. Edna Kenton, *Simon Kenton, His Life and Period 1755 -1836* (New York, N.Y.: Random House from the Doubleday edition, 1930), p 280.
17. Ted Morgan, *Wilderness at Dawn* (New York, N. Y.: Simon & Schuster, 1993), p 457.

18. *History of Hardin County* (Chicago: Warner & Beer & Co., 1883).
19. Henry Howe, *Historical Collections of Ohio,* vol. 1 (Norwalk, Ohio: Laning Printing Co., 1896), p 288.
20. ——, p 290.
21. ——.
22. *History of Miami County* (Chicago: W. H. Beers & Co., 1880), p 240.
23. *Urbana (Ohio) Daily Citizen,* Nov., 27, 1889.
24. C. W. Williamson, *History of Western Ohio and Auglaize County* (Columbus, Ohio: W. M. Linn & Sons, 1905), p 382.
25. Henry Howe, *Historical Collections of Ohio,* vol. 1 (Norwalk, Ohio: Laning Printing Co., 1896), p 659.
26. Joshua Antrim, *History of Champaign and Logan Counties From Their First Settlement* (Bellefontaine, Ohio: Press Printing Co., 1872), p 252.
27. ——.
28. W. H. Beers, *History of Darke County* (Chicago: W. H. Beers & Co., 1880), p 402.
29. Joshua Antrim, *History of Champaign and Logan Counties From Their First Settlement* (Bellefontaine, Ohio: Press Printing Co., 1872), p 295.
30. Thaddeus Gilliland, *History of Van Wert County* (Chicago: Richland & Arnold, 1906), p 139.
31. Nelson Evans, *History of Adams County* (West Union, Ohio: Emmons Stivers, 1900), p 63.
32. *Urbana (Ohio) Daily Citizen,* "Champaign Saga," 1976.
33. W. H. Beers, *History of Brown County* (Chicago: W.H. Beers & Co., 1881), p 269.
34. ——, *History of Champaign County* (Chicago: W. H. Beers & Co., 1881), p 322.
35. Eugene Roseboom & Francis Weisenburger, *A History of Ohio* (Columbus, Ohio: F. J. Heer Printing Co., 1958), p 190.
36. W. H. Beers, *History of Champaign County* (Chicago: W.H. Beers & Co., 1881), 434.
37. William A. Galloway, *Old Chillicothe* (Xenia, Ohio: Buckeye Press, 1934), p 271.
38. ——, pp 372-273.
39. Newspaper clipping from Noble County, Aug., 1880.
40. Henry Howe, *Historical Collections of Ohio,* vol. 2 (Cincinnati, Ohio: Krehbiel & Co., 1904), 467.

41. Henry Howe, *Historical Collections of Ohio*, vol. 2 (Columbus, Ohio: Krehbiel & Co., 1904), p 467.
42. Rev. David Zeisberger, *David Zeisberger's History of The Northern Indians*, ed. Archer Hubert &William Schwarz (1910; reprint, Lewisburg, PA: Wennawoods Pub., 1999).
43. W.H. Beers *History of Montgomery County* (Chicago: W.H. Beers & Co., 1881), p 7).
44. James B. Finley, *Autobiography of Rev. James B. Finley or Pioneer Life In The West* (Cincinnati, Ohio: Cranston & Curts), pp 95-96.

ET CETERA CRITTERS
Things With Wings

Chapter 13

The deep silence of the dense woods often reverberated with the clamorous gobbling of wild turkeys. Flocks numbering in the hundreds were not uncommon. Early Ohioans were known to stand in the doorways of their cabins and shoot them at will. Some of the toms, or gobblers, grew as tall as four feet and weighed as much as thirty pounds. In short bursts of speed, they could fly fifty-five miles an hour. They were nervous birds and flocks scattered at the least hint of danger. They had a tendency to stride in single file order and it was not unheard-of for a person to kill the last turkey in line and then the next and the next, working down the row until the feathered animals took alarm and flew off in different directions.

People made whistles from a small bone in a turkey's wing which produced a surprisingly turkey-like call. Those individuals who mastered the technique of blowing on the whistles often could lure the wary fowl back again to within firing range.

Somebody once alleged that turkeys would consume anything slow enough to catch and small enough to swallow. Although the assertion was somewhat overstated, turkeys did eat a wide variety of small things—among which were insects, nuts, berries and tiny green leaves. They were partial to grasshoppers and they found the pioneers' grain irresistible enough to venture into the clearings around cabins in order to eat the corn left lying on the ground. It was a widespread practice to leave the door of a corncrib open. Two and possibly more birds might be caught at the same time when they walked into the crib to eat the tempting kernels of corn.

Joshua Antrim wrote about capturing wild turkeys. "It consisted of building ordinary fence rails into a square pen, say three feet high covered with more fence rails on top, with interstices between of some four inches, making an opening on one side at the bottom of the pen large enough for a turkey to pass through it, then throw into the pen shelled corn or other cereals, trail said seed outside some distance, and very frequently a whole flock would begin on the outside trail and clean it up to the pen, and one at a time follow the leading turkey through the opening until the whole flock, large or small, would be crowded inside, and when once in they became bewildered, and had neither sense nor instinct to go out as they went in, but only attempted to escape by flying up, and were knocked back by the fence rail covering; and would either be secured in the trap until needed for use, or taken out and put into another pen and fed; and leave the trap for a new haul."[1] Some pens were constructed with a string stretched across the floor which, when tripped, dropped a door, capturing those inside.

During the winter months when a blanket of snow lay on the ground, hunters frequently tracked their game. If the soft, feathery snow was ankle or knee-deep, turkeys tired quickly when walking and predators as well as pioneers took advantage of the situation.

Reverend James Finley, a minister on the Mad River Circuit, wrote about a turkey hunt in which he had participated during the winter of 1801. The turkeys were fat that year, he said, and a great number were killed. In order to preserve them for later use, they were dressed, cut in two, salted, and hung to dry.[2]

A surveying crew of twenty-eight men in Ross County told of surviving what they termed "the starving tour." It began with a driving snow storm which lasted for four days. Trapped in the middle of the unrelenting wilderness, they had neither a tent nor a shelter of any kind. They had used all their provisions. On the third day "they luckily killed two wild turkeys, which were boiled and divided into twenty-eight parts, and devoured with great avidity, heads, feet, entrails and all."[3]

While conducting surveys in Madison County, Lucas Sullivant became aware of hostile Indians camping not far away. Advising his men to keep together, remain quiet and under no circumstances to fire a gun unless in self-defense, they resumed their work. At twilight, as turkeys began roosting in the trees, someone forgot the admonition and shot several of the birds. The Indians, now made aware of their whereabouts, lost no time attacking them. Two men were killed. The rest of the men made their escape and later regrouped. By traveling all through the night and most of the following day, they succeeded in eluding their pursuers.[4]

A history of Cuyahoga County related the story of Sarah Thorp whose husband started off through the forest to purchase badly needed supplies at a settlement twenty miles away. Mrs. Thorp and the three children, left alone, eventually ran out of food when Mr. Thorp did not return. Desperate for something to eat, they dug up roots in the woods; they emptied the straw from the bed ticking and boiled what few grains of wheat it contained; and the eight-year-old son spent fruitless hours searching for kernels of corn he thought he remembered seeing in a crack of a log in the cabin wall.

Standing in the cabin door one afternoon, watching for and fervently wishing her husband would arrive, Sarah saw a turkey alight in a tree close to the house and decided she must try to shoot it. Although she'd never hunted before, she had practiced shooting and, while not an expert marksman, did relatively well. She took down the ammunition and discovered there was enough for only a very small charge. "Carefully cleaning the barrel so as not to lose any by its sticking to the sides as it went down, she set some apart for priming and loaded the piece with the remainder, and started in pursuit of the turkey, reflecting that on her depended the lives of herself and children."

Even though she moved slowly, her approach somehow startled the turkey and it flew a short distance away. Impatiently, Sarah waited until it landed in another tree and then fluttered to the ground where it began dusting itself in the soft earth. Creeping on her hands and knees, inch by inch, from log to log, she quietly worked herself forward. Carefully, she lifted the heavy rifle. Forcing herself to remain calm, she fired. "The result was fortunate; the turkey was killed and herself and family preserved from death by her skill." It later was learned that Mr. Thorp had been murdered and never reached the settlement.[5]

When a gentleman in Montgomery County by the name of Clawson sought the services of Edmund Munger, the local blacksmith, he learned his work could not be done immediately and it might be necessary to wait for a day or two. His family was out of meat, Munger said, and he planned to hunt for turkeys. Clawson, the acclaimed turkey hunter in the community, knowing he stood a far better chance of killing turkeys than Munger, proposed that he do the hunting and Munger stay in the shop and continue his work. The proposition was accepted and Clawson left. "In the evening he returned and made good his boast for the old horse was loaded-down with twenty-one, fine, fat turkeys."[6]

Turkey meat, which ordinarily was tender and juicy, customarily was stewed in bear's oil. But—stewed in bear's oil, or roasted, or broiled, or fried—Reverend Brannen, who rode the circuit in eastern Ohio, liked turkey whatever the method of preparation. "His voice and person were huge as his appetite and he seemed proud of his eating capacity. He used to say that a turkey was an unhandy bird—rather too much for one [individual] and not quite enough for two."[7]

Although pioneers and the majority of Indians ate turkeys, there were a few tribes who abstained because they believed that consuming the flesh of timid creatures such as turkeys could propagate timidity in themselves. Turkey eggs, on the other hand, which were a pale, creamy-tan color speckled with brown and about twice the size of ordinary hen eggs, offered no such threat. Some Indians roasted those just-ready-to-hatch and thought them a special treat. The famous frontiersman, Simon Kenton, tried them and pronounced them good eating.

Turkey feathers were used in a variety of ways. Women of the Delaware Indian tribes made blankets of turkey feathers, binding them together with strands of wild hemp. In winter, these covers were said to be better protection against the cold than the heaviest woolen blankets. Primarily, however, the feathers were used for ornamental purposes.

The sharp spurs on gobblers' legs formed lethal points for Indian arrows. Turkey feet, on occasion, were utilized to make false tracks in the ground in an attempt to mislead and ambush one's enemies.

In 1884, nearly half a century after the Indian wars in Ohio were over and the last Indians had been forced onto reservations farther west, there was a much publicized dispute over domestic turkeys that resulted in a trial. The plaintiff in the trial, Louisa Runkle, brought suit against the defendant, Thomas Foster, for a brood of turkeys which was valued at $5.00. There actually appeared to be no disagreement for everybody involved, including the defendant, acknowledged that the turkeys belonged to the plaintiff. Notwithstanding, the trail proceeded. Court costs of $100.00 were assessed against the defendant, Thomas Foster, who appealed the case to

Common Pleas Court because he contended $100.00 was far too much. The higher court upheld the initial decision. By this time, court costs had risen to $150.00. A reporter who was covering the "turkey case," as it was titled, wrote that the two trials had consumed two entire days and "when it was considered that the amount involved is only $5.00, the nonsense of the suit is plainly to be seen."[8]

In the late 1700s and early 1800s as people were immigrating into the fertile land between the Ohio River and Lake Erie, over-hunting rapidly thinned the wild turkey population. As forests were converted into croplands and habitat destruction took place, the era of the wild turkey in Ohio nearly came to an end. By 1850, only a small number were left. In 1950, after careful planning, the Ohio Department of Natural Resources commenced a reestablishment program which has been perceived as highly successful. Today, the unmistakable gobbling of wild turkeys again can be heard in all eighty-eight Ohio counties.

****

Chickens were greatly valued by pioneer women and whenever possible were carried in coops or in baskets as families came to their new homes west of the Allegheny Mountains. Crates filled with chickens and ducks that swung between the rear wheels of immigrants' wagons were a common sight. Chickens were prized for the eggs they produced and for the meat they provided. As a newlywed, James Finley worked an entire day for a hen and her three half grown chicks to give to his wife.[9]

Most settlers were hospitable folks and warmly welcomed guests. They felt honored when a circuit rider stopped at their home and invariably they prepared for him the finest meal they could. This usually meant killing and cooking several of their chickens. In a relatively short time these dedicated men of God earned the name "chicken eaters." They became the brunt for many a "chicken" joke. The following anecdote is typical.

A circuit rider arrived at a remote cabin in the Ohio woods and soon thereafter heard a child crying. He located a young boy standing in back of the corncrib clutching a chicken under his arm and sobbing as if his heart would break. When asked why he was crying, the little fellow replied that between the hawks and the circuit riders, this was their last chicken and his mother had just told him to kill it in order to feed the preacher.[10]

Edwin Carlo of St. Paris, Ohio, went with a group of friends to the gold fields in California. In a letter to his parents, he said: "I suppose all is quiet again on the banks of Nettle Creek, since that preacher devoured all the fried chicken, picked his teeth with a splinter and went on his way rejoicing. I do love the ministers but sometimes I have to admit that they have a weakness for chicken."[11]

Reverend James Finley, a circuit rider himself, wrote about the Wyandot Indians and how their women took chickens along when the tribe broke into smaller groups to set-up winter camps. Reverend Finley said, "You'll often find a flock of domestic fowls which are taken on horses from the towns for the purpose of getting their eggs." To secure them from the dogs which "generally swarm around an Indian camp, the women make baskets of bark. In these they carry the chickens and once camp is set-up, they drive stakes into the ground. Sometimes they place perhaps half a dozen baskets, one above another, on one stake. From these they gather large quantities of eggs."[12]

A couple of chicken stories pertain to southerners, General "Stonewall" Jackson and General Robert E. Lee, neither of whom came from Ohio nor even fought a battle here. Information of almost any sort about these two famous officers, however, always remained of great interest to Ohioans as well as to the rest of the nation.

"Stonewall" Jackson is said to have had a pet hen. According to records, he carried her with him everywhere he went during the 1863 campaign, often making space for her in a supply wagon.[13]

Robert E. Lee, the most celebrated Confederate of them all, had a modest sized flock of chickens during one stage of the war that a well-wisher had given to him. "The staff enjoyed many meals from it. . . . One of the flock . . . deserved a better fate than the pot. Every day a lone hen laid an egg, and would choose no other spot than the general's tent, beneath his cot. . . . Every day she would walk to and fro in front of his tent, and when all was quiet walk in, find a place under his bed, and deposit her egg; then walk out with a gratified cackle. Lee always left open the tent flap for the hen, to which he became attached. For many weeks she roosted in a baggage wagon, and in the spring, when the army broke camp, she rode with the army's wagons to battle. She was to serve through several campaigns, and though the noise of battle would disturb her laying habits, she remained faithful."[14]

In 1877, a dozen years after the Union Army had defeated the Confederate Army and the country's wounds were healing, a small Ohio newspaper humorously reported that people in the community had begun referring to chicken eggs as "hen fruit."[15] The term was in effect for a period of years and then gradually faded from use.

Cockfighting, with roots in ancient Asia, was a regular sport in certain parts of Ohio. The brightly colored birds, selectively bred to produce fast, courageous roosters with a killer instinct, were profitable enterprises for promoters. Betting at these events where two gamecocks fought to the death oftentimes was heavy. Clergymen of all denominations were among the spectators, some of whom were observed staking wagers on their favorite birds.

****

Women raised domestic geese principally for their feathers. The downy breast feathers, in particular, which were softer and more pliable than either chickens' or ducks,' made the most desirable filling for comforters, bolsters and feather ticks.

Farm wives sometimes tended whole flocks of geese in order to supply so-called "feather factories." An early newspaper announced "A feather renovating establishment has been located in Millerstown."[16] Most women, however, ordinarily washed and refurbished their own feathers, considering it a part of their annual spring housecleaning.

In Brown County, a Mrs. Lyon converted her goose feathers into cash when an emergency arose. Mr. Lyon, who periodically took their produce to Cincinnati by way of the Ohio River, was detained longer than he had expected one summer. Meanwhile, their wheat had ripened and was ready to be harvested. With no money to pay a labor force, the resolute lady ripped open all their bedding, packed the feathers in a bag and, on horseback, took them to a merchant some distance away. She sold the feathers, returned home with the money and the wheat was threshed.[17]

Like cattle, horses and hogs, families marked their geese to avoid confusion with geese that belonged to other people. A relatively simple identifying mark consisted of splitting the web on either the right, the left, or possibly both feet. Clipping a small notch or two in their webbed feet was another option. Occasionally, nicks were cut in their beaks. All such markings were supposed to be registered with township clerks or with county commissioners.

Live geese oftentimes were driven to market on foot in much the same manner as larger livestock. The only problem with geese was their tender, webbed feet. Prior to a drive, it was not uncommon to force geese first to walk through a sticky substance such as tar or molasses and then over sand. The sand adhered to the tar and the coating offered some slight protection.

****

Pigeons by the thousands once flew Ohio skies. Descriptions in old history books range from "They might be seen flying in such crowds overhead as almost to darken the air, and in continuous lines for miles in length"[18] to "The air literally was filled with pigeons and the noonday light was obscured as by an eclipse."[19] John James Audubon reported watching flocks that took three days to fly past him.

At twilight, when pigeons began roosting, the crashing of tree branches and limbs breaking from the weight of the pigeons was deafening. One observer likened it to the force of a hurricane. In the spring and again in the fall during the breeding seasons, roosting areas covered as much as four acres of woods according to a report. Some viewers wrote of seeing roosting grounds extending for more than two miles in length. A single tree often contained over one hundred nests. "Dung fell in spots not unlike melting flakes of snow."[20] In the period of only one night it sometimes lay a foot high under their roosting places.

After dark, people gathered at pigeon roosts and knocked the birds from the trees with poles. By moonlight or torchlight, men clubbed them to death and with shovels scooped them into wagons. Hundreds upon hundreds were shipped in barrels to eastern markets. Nearby farmers drove their hogs to fatten on the pigeon bodies that lay under the trees.

Peter Rockel of Clark County remembered going with his brother to an area near their home in Tremont City they knew as Pigeon Swamp . With lanterns and long flails, the two boys started out. They could hear the constant cooing and commotion among the birds long before they reached the place. Once there, the bright glare from the lanterns bewildered the birds who flew around the boys in a frenzied, swirling mass. Fearing they would be smothered, the brothers dropped their lanterns and flails on the ground and fled in terror.[21]

The following excerpt was copied from a paper that was prepared for a circa 1943 county historical society meeting:

"To catch the large flocks of pigeons that many years ago fluttered over the Ohio forests, the men used decoys in somewhat the same way that duck hunters now use decoys except that the pigeon-decoys were alive.

"After first sewing the pigeons' eyelids shut, the men in these parts, would then tie a cord to their legs and place them on shelves built high in the air. This cord ran through a hole in the center of the small platform to the hands of the hunters who could thus control their decoys and keep them from flying away. Perched on these elevated stands the pigeons would flutter their wings in an effort to take to the air, and in so doing would attract other pigeons.

"Upon arriving, the flocks would swoop down for a lunch already prepared for them by the hunters and while eating, a large net would fall over and trap them."

The writer concluded the article by explaining the word "stool-pigeon" originated from the practice of placing decoy-pigeons.[22]

Although passenger pigeons once were seen by the millions, the species were killed in such appalling numbers they eventually became extinct. Scientists believe that the last passenger pigeon in existence—who fondly was known as "Martha"—died in 1914 in the Zoological Gardens in Cincinnati, Ohio.

****

Along the banks of the Great Miami River, the Scioto River, and the Ohio River, some of the first pioneers reported seeing large flocks of parrots darting in and out through the trees. The birds, their bodies about the size of doves, were green; their heads and the tips of their wings were a brilliant red and in the sunlight they fairly glowed. Their squawking was raucous and unpleasant.

Colonel Donn Piatt, the Logan County gentleman who trained his dog, Frank, to summon and then to "bark" each family member to breakfast every morning, kept a parrot he named "Gray Poll." In a book written about Colonel Piatt's life, a friend related an incident involving the parrot.

"Gray Poll, on her perch in a rear court, had observed that whenever anyone cried 'Sic 'em!,' the dogs rushed from their kennels in a pack and tore off barking furiously. This proceeding gave Poll delight and she never failed to take up the cry, screaming, 'Sic 'em!' in her shrillest tone. She also had learned to know as well as did the dogs the meaning of 'Get out!' It occurred to Gray Poll, one day, that she might have a little harmless fun with the dogs lying in the sun, and she suddenly set up a cry of 'sic 'em!' The dogs were up in an instant dashing about the lawn, and Poll laughed in high glee at the poor brutes' vain effort to find something to attack. Poll kept up the sport until one of the dogs discovered her trick and pulling her from her perch the whole pack set upon her and would soon have ended her life but for a sudden inspiration that came upon the poor bird to cry 'Get out, get out, ____

you, get out!' The dogs made off, and poor Poll, hopping painfully back upon her perch, disconsolately surveyed her torn plumage, sought an easier position for her broken bones and solemnly observed to herself; 'Gray Poll talks too d_____ much.' "[23]

****

Hawks, ever present in Ohio, were considered an abomination by people who raised chickens and other types of fowl for they could decimate a brood of new chicks, ducklings, or goslings in short order.

When a fifty cent bounty for each hawk head was offered in Trumbull County, an industrious young fellow began breeding them. As they came of age, he cut off their heads and delivered them to the proper authorities in Warren where he collected his money.[24]

At cross purposes were the sportsmen who tried to revive the use of certain kinds of hawks for hunting and advertised for their eggs at forty dollars apiece. A few courageous men are said to have braved the piercing whistles, screams, and chattering calls of the parent birds and gathered eggs from the nests by lowering themselves over the side of a cliff. Some of the hawks attacked the interlopers, causing painful cuts with their sharp talons and beaks.[25]

****

Sparrows, brought to America from Europe about 1850, swiftly increased in numbers. They apparently had found their way to the State of Ohio twenty-nine years later because the following item appeared in an 1879 newspaper:

"Andrew J. Green noticed the strange actions of a large flock of sparrows, hovering down low over the sidewalk, flying speedily hither and thither in great excitement. As he approached them and when in their midst, they evinced no fear of his presence and instead of flying away as he walked among them, they pressed around him in great numbers and attacked him with their sharp bills with great fury. At first he dismissed the attack as a trivial matter and attempted to brush them away with his hand but they persisted in their attack, darting, chattering and piercing him with their sharp bills. Their immense number and persistent charge threw him to the ground. Now alarmed, he struggled to his feet and covered his eyes with his hand while trying with the other hand to wrest a stake from a fence to defend himself. He tried to swat them away and again was thrown to the ground. Now really terrified, he pulled his coat over his face, regained his feet and ran as hard as he could. His hands were bleeding."[26]

****

While there still is debate concerning the origin of honeybees, some experts believe they are not indigenous to the North American continent. Ships' manifest records indicate that honeybees may have been imported with some of the first colonists in 1638. Native Americans were fascinated with the winged insects that produced such a delectable, sweet-tasting substance and called them "English flies." As bees rapidly spread northward, southward, and westward, "the Indians considered them the harbinger of the white man, as the whites do the buffalo and deer of the Indian, and note that as the larger game retires the bee advances."[27]

According to Henry Howe, author of *Historical Collections of Ohio,* honeybees were the "luxury of the wilderness." He described how people located bee trees in the forest by attempting to follow a bee in flight. Bees, who could fly as fast as fifteen miles per hour, were not especially easy to keep in sight. "The hunters run after them with head erect and eyes aloft, frequently stumbling over obstacles at their feet; in this manner they track bees to their individual colonies, mark the trees, and seek for more. They dare not cut down the trees until fully prepared to take away the honey, for the bears, skunks, raccoons and possums have sweet teeth and would soon devour any honey within their reach."[28]

It was the accepted practice to mark a bee-tree in order to claim ownership; however, that did not necessarily guarantee possession. A gentleman in Van Wert County discovered a bee-tree and accordingly marked it by carving his name on the trunk. A day or two later, he discovered that somebody had removed his name and carved "Scott," the name of a querulous man who lived in the community. It was decided to disregard the under-handed attempt to lay claim to the honey and a group of the initial owner's friends worked with him to cut down the tree. Fearing Scott would hear them chopping, they wrapped a chain tightly around the trunk to deaden the sound. The tree was cut and two large buckets of honey collected when they heard someone coming. Assuming it was Scott, they quietly hurried out of sight, hoping the disagreeable man never would learn who had taken part in the effort.[29]

In 1844, a brutal murder was committed because of bees. "A swarm of bees was stolen from Felty Jacobs. He accused William Shamlin, a person of questionable habits, of being the thief. This maddened the latter. At a favorable moment he struck Felty, crushing his skull and killing him almost instantly. Shamlin was arrested and confined in the county jail but escaped by the assistance of his wife, who had obtained employment at the jail."[30]

Anna Bently, a Quaker who lived with her husband and their six children in Columbiana County, maintained a bee hive. In a letter to her parents in Maryland, dated Aug. 27, 1826, she wrote that someone had stolen the hive. Worst of all, she lamented, It "was nearly full of honey. . ." Then, almost as an afterthought, she added to the sentence, "botheration to them."[31]

Wooden bee hives and bee skeps, which were receptacles made of coils of straw or dried grass, were kept by many early Ohioans to house bees. An equally effective but simpler method of corralling bees was to use a section of a hollow log. Stood upright, with holes bored at the bottom for the bees to enter and leave and a removable lid at the top, the honey could be harvested. Honey sometimes was removed at night to lessen the chance of being stung.

Boys delighted in accepting challenges from their companions and did battle with honeybees during daylight hours to try to take the honey. Armed with a leafy branch or a handful of weeds to swat the bees away, they would fight their way through to the hive and scoop up the honey.

The sweet-tasting stuff was used to season the often bland food that people ate. It also was used to preserve hams and fruit. Beeswax, the substance from which honeybees made the cells of their combs, was widely used in medicine; it was used to bind wounds; it sometimes was used to mold candles; and leather, rubbed well with beeswax, was found to be waterproof.

A man and his wife in Crawford County befriended an Indian family who one day cut down a bee-tree, gathered the honeycomb and gave the settlers a large piece of it. "The man was quite overcome by the generosity of the Indian" and said, "The Indian was gentle in peace, while desperate and brave in war."[32]

Robert Price, in his book, *Johnny Appleseed: Man and Myth,* wrote that Johnny "would never touch tea, coffee, or tobacco, because when he got to the next world, he could not have them and so would not cultivate a taste for them here. But milk and honey were different. . . . This is considered heavenly food." While wild honey almost literally flowed at times in the Ohio woods and the settlers helped themselves freely, Chapman, if he found a bee tree, always looked carefully to see whether the insects had sufficient store for the winter before he touched the comb."[33]

****

The April 15, 1880, *St. Paris Examiner* published an item about two men and a couple of wasps: "Two men each put up 5 dollars apiece on a wager that one could hold a wasp in his hand longer than the other could, and the fellow who rubbed chloroform on his hand expected to win, but the other fellow happened to know that male wasps don't sting, and got one of that sex, and they grabbed their wasps and sat and smiled at each other while the crowd wondered, until the chloroform had evaporated, and then the fellow who used it suddenly let go of his wasp, and let the audience into the secret of how to swear
the shingles off the roof." (Note how the copy is one l-o-n-g sentence!)

H. Z. Williams, in *History of Preble County,* wrote that in 1806 the Silas Dooley family lived near Big Sulphur Springs, Ohio. They became acquainted with an Indian and his family who set-up a camp not far from their cabin. Oftentimes, the Dooleys gave the family pumpkins and other produce from their garden. One day the Indian killed a bear and a deer and offered to share the meat. Silas tied his portion in a bundle and slung it across his shoulders. Bent nearly double under the weight and with eyes focused on the ground, he asked his Indian friend to guide him back to the cabin. The Indian, highly amused at the manner in which Silas was carrying the meat, decided to play a prank at the white man's expense. He led Silas under a low-hanging hornets nest and as they tumbled out of their paper-thin nest, he adroitly leaped aside and ran a safe distance away where he stood watching, overcome with laughter. Mercilessly, the hornets stung every exposed part of the settler's body but, determined not to drop the meat, he staggered on threatening vengeance with every step.[34]

John Chapman, aka Johnny Appleseed, while assisting a group of settlers clear a road through the woods, "in the course of their work accidentally destroyed a hornets' nest. One of the angry insects soon found a lodgment under Johnny's coffee-sack cloak, but although it stung him repeatedly he removed it with the greatest gentleness. The men who were present laughingly asked him why he did not kill it. To which he gravely replied that 'It would not be right to kill the poor thing, for it did not intend to hurt me.' "[35]

****

Horseflies, the vernacular name for gadflies, or breeze flies, were a menace to farm animals. Livestock were made miserable by them. An elderly Clark County man was quoted as saying horseflies, "which grew almost to the size of a mouse, would set the horses and oxen frantic with its terrible sting."[36] Pioneers heading to northeastern Ohio in 1799 with yokes of oxen, reported. . . "their oxen were tormented and rendered almost unmanageable by immense swarms of large flies, which displayed such skill in the science of phlebotomy, that, in a short time, they drew out a large share of blood belonging to these animals; the flies actually killed one of the oxen."[37]

The ordinary housefly was a great annoyance to man and beast alike. A visitor in Sandusky, Ohio, recorded his thoughts about the flies he saw there. "I was struck with one singularity—the air was filled and every sunny wall or building was covered with myriads of a disgusting fly about an inch long, with large wings and feelers. . . . They find their way into the houses and infest everything, even the table where we dined swarmed with them."[38]

A stagecoach traveler once remarked how pleasant it was when he stopped in a central Ohio town and the innkeeper's wife not only served an appetizing meal but also "stood by me fanning the dishes to keep off the flies."[39]

Many early kitchens were painted blue for it was the mistaken belief that flies were repelled by the color blue.

The following jingle was found inserted between the pages in an old book that once belonged to Catherine Norman, a nineteenth century maiden lady in west central Ohio.

> Fly, Flee, Flaw, Flue
> A fly and a flea in a flue
> Were imprisoned. Now what could they do?
> Said the fly, "Let us flee!"
> "Let us fly!" said the flea.
> So they flew through a flaw in the flue.
>
> Puck[40]

****

Mosquitoes, one of mankind's most deadly enemies, were diminutive in size but massive in number. Each insect, armed with two minuscule serrated knives for stabbing, two probes for spreading the incision apart, and a long tube for sucking blood, was a menace to humans and livestock alike.

In 1785, when the United States Congress ordered the Northwest Territory surveyed, surveyors began their task of determining and fixing boundaries. While working in the area of present-day Ohio, surveyors and their crews encountered obstacles of all sorts—some life threatening; some merely a source of irritation. Squatters pulled up survey stakes and shot at the men from behind trees. Some of the men drowned, some froze to death, some were scalped. There were those who claimed that if the climate didn't get them, the Indians would. Cases of poison ivy were common. The mosquitoes were insufferable, especially when the groups passed through swamps and marshy areas—the very places that afforded ideal breeding grounds for mosquitoes. Some of the men insisted that the two-winged insects would engorge themselves on human blood until, if left undisturbed, would swell to four times their ordinary size and then fall off and actually burst. No one could sleep at night without the benefit of the smoke from the fires which helped to keep the insects at bay.

Thomas Worthington learned first hand how brutal mosquitoes could be at night in a swamp. As a young man he once agreed to help paddle a canoe up the Scioto River to deliver mail. When the men camped for the night on the bank of the river, Worthington went for a walk and lost his way. Overtaken by darkness, he found himself in a swamp and "was half devoured by the myriads of mosquitoes." The following morning, his companions waited as long as they dared but were compelled to continue without him. Worthington, meantime, had struck out at daybreak and, locating the river again, waited, hoping the men had not already passed. To his immense relief he saw the canoe as it came into sight and, hailing them, rejoined the party.[41]

Mosquito Lake, in Champaign County, aptly named for the legions of mosquitoes that annually infested the vicinity "really was a pretty little lake if one stood on the surrounding hills and looked down at it," once mused a local resident.[42] Except for the splendid ice skating it offered in the wintertime, little else of a favorable nature could be said about the place.

Mosquito Creek, in Trumbull County, derived its name in an equally inauspicious manner. In 1818, according to Henry Howe in *Historical Collections of Ohio*, there were few wells or cisterns in Trumbull County and women in the region of Mosquito Creek had to endure the bothersome insects as they did the family's washing in its water.

Settlers oftentimes smeared animal grease over the exposed parts of their bodies in an effort to protect themselves from the painful jabs. Families laid damp chips of wood over live coals to make smudges which produced a thick smoke. Placed in front of cabin doors, and sometimes inside the dwellings, it brought relief to people in the summer evenings. "The remedy was severe but preferable to the stinging of the assiduous mosquitoes."[43]

Farmers burned green-timber fires to protect their livestock from the insects. The animals soon learned of their own accord to seek out and stand in the smoke in order to be free of the torturous pests. Bears knew to climb high in the trees where the cool breezes helped keep insects away.

One of the histories of Putnam County told about Indian Tom, a "bad Indian," according to the writer. "In the spring of 1833, he stole a pony from some of his tribe. They tried him for stealing, found him guilty, took him from camp, divested him of his clothing, laid him on his back, tied him to a stake, and left him to remain all night, subject to the attack of the innumerable hosts of mosquitoes and gnats. I saw Tom the next morning; he was a fearful looking object. He looked as though every pore of his skin had been penetrated by the insects. I sympathized with him, notwithstanding I knew he was a thief. After Tom was released they procured whiskey and the whole tribe, except Pe Donqet, the chief, got drunk and had a general spree lasting two days."[44]

Knox County, in central Ohio, was the scene of an unusual contest. Antagonists John Daily and Alexander

Darling, each bet a quart of whiskey "about the ability of a man to withstand black ants, ticks and all else, except 'gallinippers' (mosquitoes), without flinching." Daily chose to try first, and stripping off his shirt, lay on a sandy spot thick with black ants. Darling, as a condition of the wager, agreed to keep all gallinippers away from Daily. After lying for several minutes without so much as twitching while ants crawled over him, Darling began letting the mosquitoes bite Daily. Still, the man didn't flinch. Then Daily, who was determined to win the bet, whispered to one of the spectators to bring a live coal from the fire and place it on his back. This was done and for a full fifteen minutes he endured the fearful pain as the hot coal seared his flesh. He moved not a muscle. Throwing up his hands, Alexander Darling conceded the match without venturing to surpass the feat. He decided a quart of whiskey wasn't worth all that his opponent had suffered. Daily was welcome to it.[45]

A story that illustrated Johnny Appleseed's unconventional lifestyle and his kindness to all living things involved mosquitoes. "On one cool autumnal night, whilst lying by his camp fire in the woods, he observed that mosquitoes flew in the blaze and were burnt. Johnny thereupon brought water and quenched the fire, afterwards saying: 'God forbid I should build a fire for my comfort, that should be the means of destroying any of His creatures.' "[46]

In 1862, while the Civil War was raging, a group of twenty-two men, most of whom were from the 21st Regiment Ohio Volunteers, joined ranks in a daring raid deep into enemy territory to capture the Confederate railroad locomotive known as the "General." On the morning of April 12th, the locomotive with several cars successfully was captured in Marietta, Georgia, but due to lack of fuel and water, the train coasted to a stop well within Confederate lines. All the men eventually were taken prisoner

Eight of the men ultimately were executed as spies. Seven were Ohioans and all seven of them posthumously were awarded the first Congressional Medals of Honor ever to be presented.

William Pittenger, one of the participants in the raid, wrote about the experience in a book called *Capturing a Locomotive*. In his book, Pittenger told about the successful escape of two of the men from an Atlanta prison. These two, instead of heading north toward Union lines, traveled south on the Chattahoochee River. The swamps were horrendous and the insects even worse. One of the men said, "Besides the torments of hunger, our nights were made almost unendurable by the swarms of blood-thirsty mosquitoes, which came upon us in clouds. I did think that I had learned considerable about mosquitoes in my boyhood in the Black Swamp of Northwestern Ohio, but for numbers, vocal powers, and ferocity, I will trot the Chattahoochee Swamp fellows out against any others I have ever met up with. The ragged clothing, which yet clung to our backs, did not much more than half cover us. To protect ourselves from the pests, we thatched our bodies all over with great skeins of moss, and two more comical-looking beings than we were, thus rigged out, it would be hard to find, but it baffled the bills of our tormentors." At another point, one of the men remarked: "It did seem as if the mosquitoes would carry us away piecemeal."

After an appalling trip, the two escapees reached the Gulf of Mexico and rowed alongside a Union ship, one of the Federal blockading squadron anchored in harbor at Appalachicala Bay. There was, at first sight, disbelief that the weird-looking creatures actually were northern soldiers. Once convinced, the bedraggled men were hustled aboard. As they were given food and the medical officer tended to their emaciated bodies, others crowded around to hear their graphic account.[47]

Both before the Civil War began and after it had ended, many Ohioans' favorite pastime was attending camp meetings. From mid-summer to early fall when camp meetings were in full swing, local newspapers announced the locations and dates. Reporters printed human interest stories regarding the events. An amusing quip inserted between articles in one of the papers read: "The mosquitoes are eagerly comparing camp meeting advertisements."[48]

Bloodletting, the process of opening a vein or artery for the treatment of disease, was based on an ancient Greek belief that the body consisted of four humors—blood, phlegm, black bile, and yellow bile—and any imbalance of the four could result in illness. Bleeding a pint or more of a patient's blood was thought to restore the balance. It became a common method of "curing" nearly all disease and was practiced throughout much of the world.

One eighteenth century doctor concluded that the human body contained twenty-five pounds of blood, more than twice the actual amount, and he advocated the removal of as much as four-fifths of it. Bleeding to the point of unconsciousness was perceived to be the most beneficial.

Bloodletting was performed by doctors, barbers, and, often, individuals themselves, using a lancet, a surgical instrument with small, sharp blades. Circuit riders were known to carry a lancet with them as they traveled through the untamed wilderness and bleed themselves when they felt it was advisable.

In Wood County, Ohio, there were residents who said there was no need for a doctor to bleed them for the mosquitoes in the area did all the bleeding that was necessary.

Because the dislike of mosquitoes was universal, it almost was with a sense of satisfaction that people read the following item in an 1870 *Mechanicsburg Review*: "Everyone will rejoice to learn that a distinguished New Jersey doctor has found that mosquitoes have parasites which give those annoying insects a great deal of trouble. A number of captive mosquitoes were examined under a microscope, and they were all found to be positively lousy, each having parasites upon its body which were easily to be seen under a glass of low magnifying power. The mosquitoes seemed to be as lively under their torture as human beings are under the sting of the mosquito. It is quite a comfort to know that they, too, can suffer."[49]

Chapter 13: ET CETERA CRITTERS; Things With Wings

1. Joshua Antrim, *History of Champaign and Logan Counties From Their First Settlement* (Bellefontaine, Ohio: Press Printing Co., 1872), p 21.
2. James B. Finley, *Autobiography of Rev. James B. Finley or Pioneer Life in the West* (Cincinnati, Ohio: Cranston & Curts, 1853), p 152.
3. Henry Howe, *Historical Collections of Ohio* vol. 2 (Cincinnati, Ohio: Krehbiel & Co., 1904), p 505.
4. W. H. Beers, *History of Madison County* (Chicago: W. H. Beers & Co., 1883), p 307.
5. Henry Howe, *Historical Collections of Ohio,* vol. 1 (Norwalk, Ohio: Laning Printing Co., 1896), pp 527-528.
6. W. H. Beers, *History of Montgomery County* (Chicago: W. H. Beers & Co., 1882) p 7.
7. Henry Howe, *Historical Collections of Ohio,* vol. 1 (Norwalk, Ohio: Laning Printing Co., 1896), p 892.
8. *Urbana (Ohio) Citizen and Gazette*, January 3, 1884.
9. Samuel H. Stille, *Ohio Builds A Nation* (Chicago: Arlendale Book House, 1962), pp 190-191.
10. Henry Howe, *Historical Collections of Ohio,* vol. 1 (Norwalk, Ohio: Laning Printing Co., 1896), p 892.
11. An 1870 letter from Edwin Carlo, St. Paris, Ohio.
12. James B. Finley, *Life Among The Indians* (Cincinnati, Ohio: Hitchcock & Walden, 1857), p 296.
13. Tony Horwitz, *Confederates in The Attic* (New York, N.Y.: Pantheon Books, 1999) p 24.
14. Burke Davis, *Gray Fox* (New York, N.Y.: Holt, Rinehart & Winston, 1956), p 174.
15. *(St. Paris, Ohio) New Era*, 1877.
16. *St. Paris, (Ohio) Era-Dispatch*, November 27, 1890.
17. W. H. Beers, *History of Brown County* (Chicago: W. H. Beers & Co., 1883), p 372.
18. Joshua Antrim, *History of Champaign and Logan Counties, From Their First Settlement* (Bellefontaine, Ohio: Press Printing Co., 1872), p 241.
19. Walter Havighurst, *River To The West* (New York, N.Y.: Penquin Group., 1948), p 153.
20. ———.
21. Orton G. Rust, *History of West Central Ohio* (Indianapolis, IN: Historical Pub. Co., 1934), p 352.

22. R. H. Ross, a paper prepared for a Champaign County Historical Society meeting (n. d., possibly circa 1943).
23. C. G. Miller, *Donn Piatt: His Work And His Ways* (Cincinnati, Ohio: Robert Clarke & Co., 1893), pp 296-297.
24. *(Urbana, Ohio) Daily News,* May 27, 1882
25. ———.
26. *Urbana (Ohio) Citizen and Gazette,* February 20, 1879.
27. Henry Howe, *Historical Collections of Ohio,* vol. 1 (Norwalk, Ohio: Laning Printing Co., 1896), p 491.
28. ———.
29. Thaddeus Gilliland, *History of Van Wert County* (Chicago: Richland & Arnold, 1906), p 149.
30. W. H. Beers, *History of Champaign County* (Chicago: W. H. Beers & Co., 1881), pp 397-398.
31. *The Ohio Frontier: An Anthology of Early Writings,* ed Emily Foster (Lexington, KY: University of Kentucky Press, 1996), p 165.
32. Henry Howe, *Historical Collections of Ohio,* vol. 1 (Norwalk, Ohio: Laning Printing Co., 1896), p 491.
33. Robert Price, *Johnny Appleseed: Man and Myth* (Urbana, Ohio: Urbana University, 1956), p 170.
34. *History of Preble County* (Cleveland, Ohio: H. Z. Williams & Bros., 1881), p 177.
35. Joshua Antrim, *History of Champaign and Logan Counties From Their First Settlement* (Bellefontaine, Ohio: Press Printing Co., 1872), p 157.
36. W. H. Beers, *History of Clark County* (Chicago: W. H. Beers & Co., 1881), p 674.
37. Henry Howe, *Historical Collections of Ohio,* vol. 2 (Cincinnati, Ohio: Krehbiel & Co.,1904), p 627.
38. Cyrus P. Bradley, *"Journal of Cyrus P. Bradley, A Trip Through Ohio and Michigan in 1835"* (Columbus, Ohio: Ohio Archaeological and Historical Quarterly, 1906,) p 251.
39. Henry Howe, *Historical Collections of Ohio,* vol. 2 (Cincinnati, Ohio: Krehbiel & Co., 1904), p 196.
40. Clipping inserted between pages of a book that once belonged to Miss Catherine Norman, St. Paris, Ohio.
41. A. B. Sears, *Thomas Worthington* (Columbus, Ohio: The Ohio State University Press, 1958), p 14.
42. Ben Riker, *Pony Wagon Town* (Indianapolis, IN: Bobbs-Merrill Co., 1948), p 229.
43. *History of Logan County and Ohio* (Chicago: O. L. Baskin, 1880), p 325.

44. Henry Howe, *Historical Collections of Ohio,* vol. 2 (Cincinnati, Ohio: Krehbiel & Co., 1904), p 469.
45. A. Banning Norton, *History of Knox County* (Columbus, Ohio: R. Nevins Printer, 1862), p 315.
46. D. B. Beardsley, *History of Hancock County* (Springfield, Ohio: Public Printing, 1881), p 135.
47. William Pittenger, *Capturing a Locomotive* (Philadelphia, PA: J. B. Lippincott & Co., 1881), p 9-25.
48. *(St. Paris, Ohio) New Era,* August 16, 1877.
49. *Mechanicsburg (Ohio) Review*, September 1, 1870.7

AN ERA ENDS

Chapter 14

The era of the early Ohioan is over and its past forever is relegated to family records and history books. It was a period when multitudes flocked to the country which lay nearly midway between the Atlantic Ocean and the Mississippi River. It was an exciting time, for during that stage of development it became a state—Ohio—the very first state to be carved from the vast Northwest Territory.

Tales of the beautiful, primeval forests and the flat, grass covered prairies beckoned to eager immigrants, men and women who had dreams of making homes in the virgin land. Equally as exciting were the reports of the unbelievable numbers of wild animals which stirred the blood of hunters and trappers. For whatever their reason, people came. Although they were willing to take the risks as they headed for Ohio, probably few had any idea of the difficulties and dangers that loomed ahead.

Many did not survive. They drowned in rivers and streams, they were crushed under wagons, and they died of diseases and lack of medical attention. Some were killed by Indians. Always, there were the wild creatures; many were predators that crept, crawled, sprang, stalked, and lunged, maiming people and invading barn lots to devour livestock.

Some of the newcomers prospered. Others merely eked out a living. Some moved on westward and established homes there. Those who remained in Ohio built houses, many of which were constructed of logs, the timber cut from their own property. Persons who were financially able erected houses of brick, brick usually molded and fired right on the premises. Schoolhouses were built because farsighted inhabitants perceived the value of an education.

As the years piled one on top of another, rapid changes occurred. Children in the mid-1800s no longer considered themselves "pioneers" or even "early settlers." They listened though to stories about the wild animals and envisioned the hardships their parents and grandparents had endured while clearing the land and rearing their families. They learned the frontier had not been a place for the fainthearted. In their classrooms, youngsters sang songs that reflected the by-gone days. One of the songs was titled "New Country Song:"

>This wilderness was our abode
>   some sixty years ago,
>And if good meat we wished to eat
>   We shot the buck or doe;
>For fish we used the hook and line,
>   We pounded corn to make it fine,
>On Johnny cake our ladies dined
>   In this new country.
>
>Our paths were winding through the wood
>   Where the savage often trod;
>They were not wide, nor scarce a guide,
>   But all the ones we had.
>Our houses, too, were made of wood
>   Rolled into squares and chinked with mud;
>If the bark was tied the roof was good
>   In this new country.
>
>The Indians often made us feel
>   That there was danger nigh;
>The shaggy bear was often where
>   The pigs were in the sty.
>The rattlesnakes our children dread:
>   The fearful mothers often said,

"Some beasts of prey will take away
    My babe, in this new country.

Our occupation was to make
    The lofty forests bow,
With axes good we chopped the wood,
    For all we well knew how.
We cleared the land for rye and wheat
    For strangers and ourselves to eat;
From maple trees we drew our sweet
    In this new country.

For Statesmen here we had no need
    Upon our toil to feed:
For doctor's pills we had no bills,
    No tithes to pay the priests.
Our healths needed no repairs;
    No pious man forgot his prayers.
Who could fee a lawyer here
    In this new country?

Of deerskins we made moccasins
    To wear upon our feet:
A checkered shirt was thought no hurt
    Good company to keep.
And if a visit we should take
    On a winter's night or winter's day,
The oxen drew our ladies' sleighs
    In this new country.

Our little thorns bore apples on
    When the blackberries were gone;
The sour grapes we used to take
    When frosty nights came on.

> For wintergreen our girls would stray—
>   For butternuts boys climbed the trees,
> And sassafras root was our latest tea
>   In this new country.
>
> And if our boys they wished to court
>   The girls were never shy,
> For Cupid drew his bow, and
>   Slew the parties down so sly!
> In fine, we were a happy band,
>   And wished not from a foreign land
> Nor craved the gold from India's strand
>   In this new country."

from *St. Paris (Ohio) New Era*, August 29, 1878.

Although the period of the Ohio pioneer has run its course, traces linger. Names, for instance, may have the power to evoke a sense of former days and recall the kinds of creatures our ancestors encountered. Maps, both old and recent ones, reveal many of the names that still prevail—names like:

- Snake Road in Montgomery County and Snake Run in Fairfield County.
- Snake Hollow School in Fairfield County.
- Black Snake Creek in Champaign County and Black Snake Road in Licking County.
- Rattlesnake Creek in Delaware, Fayette and Highland counties.
- Rattlesnake Island in Lake Erie and Rattlesnake Knob in Ross County.
- Wolf Creek in Medina, Montgomery, Morgan, Ottawa, Seneca, Stark, Summit, and Washington counties.

Wolf Ditch in Van Wert County and Wolf Run in Pickaway County.
Panther Creek in Clark, Darke, Hardin, Miami, and Shelby counties.
Little Panther Creek in Darke and Miami counties.
Panther Run in Butler and Licking counties.
Bob Tail Pike in Champaign County.
Wildcat Creek in Hardin and Paulding counties and Wildcat Pike in Marion County.
Bear Creek in Clermont, Fulton, Mercer, Montgomery, Putnam, Williams, and Wood counties.
Little Bear Creek in Columbiana County.
Bear Run in Richland County and Bear Run Church in Perry County.
Bear Hollow Road in Licking County.
Deer Creek in Fayette, Fulton, Madison, Pickaway, Putnam, Ross and Trumbull counties
Buck Creek in Champaign, Clark, and Lorain counties
Buck Run in Allen, Hancock and Union counties.
Buckrun Creek in Hancock County and Buckskin Creek in Defiance County.
Deersville, a town in Harrison County.
Elk Creek in Butler County, Elks Run in Muskingum County, and Elk Fork in Noble County.
Elkhorn Creek in Carroll and Jefferson counties.
Buffalo, a town in Guernsey County.
Buffalo Creek in Noble County and Bullskin Creek in Clermont and Lawrence counties.
Cowpath Road in Champaign County and Little Cow Run in Washington County.
Horse Run in Paulding County.
Black Horse, a town in Portage County.

Hog Creek in Allen, Champaign and Hardin counties and Little Hog Creek in Allen Count
Hog Creek Run in Jackson County and Hog Run in Miami and Paulding counties.
Hog Creek Ditch in Hardin County and Hogback Run in Defiance County.
Hogtrail Run in Jackson County and Hogpath Pike in Miami County.
Swine Creek in Trumbull County.
Sheep Run in Brown County and Mutton Ridge Road in Muskingum County.
Pigtail Alley in Amsden and in Westville, Ohio.
Dog Creek in Van Wert County and Dog Run in Paulding County.
Dogtown Road in Huron County and Dog Leg Road in Champaign County.
Polecat Road in Green and Miami counties.
Beavercreek, a town in Greene County.
Beaver Creek in Clark, Geauga, Henry, Loraine, Pike, Seneca, Williams, and Wood counties.
Little Beaver Creek in Henry and Wood counties.
Beaverdam Creek in Tuscarawas County.
Beaver Run in Allen, Coshocton, Marion, and Paulding counties.
Little Beaver Creek and Beaver Lake in Columbiana County.
Opposum Creek in Monroe County and Opposum Run in Miami County.
Possum Run in Richland County and Possum Road in Clark County
Raccoon Creek in Gallia, Jackson, Licking and Vinton counties.
Little Raccoon Creek in Jackson, Sandusky and Vinton counties.

Raccoon Run in Fairfield and Licking counties.
Raccoon Road in Gallia County and Raccoon Valley Road in Licking County.
Raccoon Island in Gallia County.
Coon Creek in Sandusky and Williams counties.
Coonpath Road in Fairfield County.
Turkeyfoot Creek in Fulton and Henry counties and Turkeyfoot Corner in Ashtabula County.
Turkeyridge Road in Knox County.
Turkey Creek in Adams County.
Turkey Run in Madison, Morrow and Perry counties and Turkeyhen Run in Washington County.
Duck Creek in Trumbull and Washington counties and Duck Run in Putnam County.
Chicken Foot Crossing in Clark County.
Goose Creek in Hocking, Holmes, and Warren counties.
Goose Run Road in Athens and Brown counties.
Pigeon Creek in Jackson, Summit and Vinton counties.
Pigeon Run in Allen and Stark counties.
Owl Creek in Auglaize, Fulton, Hamilton, Henry, Seneca and Williams counties.
Eagle Creek in Hamilton, Hancock, Paulding, Portage, and Williams counties.
Eagle City in Clark County.
Crow Creek in Medina County.
Bee Run in Marion and Warren counties.
Mosquito Creek in Champaign and Shelby counties.

The end.

Map of Ohio
with all 88 counties

BIBLIOGRAPHY/

Alder, Henry Clay. *History of Jonathan Alder: His Captivity and Life With The Indians.* 1870. Reprint. Edited by Larry L. Nelson. Akron, Ohio: University of Akron Press, 1999.
Alderton, David. *Foxes, Wolves: Wild Dogs of The World.* N. Y.: Sterling Pub. Co.,1994.
——*Wild Cats of The World.* New York: Sterling Pub. Co., 1993.
Aldrich, Lewis Cass. *History of Erie County.* Syracuse, N.Y.: Mason & Co., 1889.
Antrim, Joshua. *History of Champaign and Logan Counties, From Their First Settlement.* Bellefontaine, Ohio: Press Printing Co., 1872.
Annual Report of The Secretary of The State of Ohio, 1885. Columbus, Ohio: R. Nevins Printer, 1886.
Banta, R. E. *The Ohio.* New York: Rinehart & Co., 1949.
Bauer, Erwin. *Bears in Their World.* N.Y.: Outdoor Life Books, 1985.
Beardsley, D. B. *History of Hancock County.* Springfield, Ohio: Public Printing, 1881.
Belote, Julianne. *The Compleat American Housewife.* Benicia, CA: Nitty Gritty Productions, 1974.
Belue, Ted. *The Long Hunt.* Mechanicsburg, PA: Stackpole Books, 1996.
Beers, W. H., *History of Allan County*, 1885, W. H. Beers & Co., Chicago
——. *History of Brown County.* 1883.
——. *History of Champaign County.* 1881.
——. *History of Clark County.* 1881.
——. *History of Darke County.* 1880.
——. *History of Madison County.* 1883.
——. *History of Miami County.* 1880.
——. *History of Montgomery County.* 1882.
Bone, J. H. A. *Adventures of Matthew Brayton The Indian Captive.* Cleveland, Ohio: Harold Office Printers, 1860.
Briggs, Hilton M. *Modern Breeds of Livestock.* N.Y.: McMillan Co., 1958.
Broadstone, M. A. *History of Greene County.* Indianapolis, IN: Bowan & Co., 1918.
Chase, A. W. *Dr. Chase's Recipes....* Detroit, MI: F. B. Dickerson, 1867.
Christopher Gist's Journals With Historical, Geographical and Ethnological Notes and Biographies of His Contemporaries. 1750. Reprint, Edited by William Darlingrton. Pittsburgh, PA: J. R. Weldin & Co., 1893.

Crain, Ray. *Land Beyond the Mountains.* Urbana, Ohio: Main Graphics, 1994.
——. *Long Green Valley.* Urbana, Ohio: Main Graphics & Associates, 1996.
——. *Simon Kenton, The Great Frontiersman.* Urbana, Ohio: Main Graphics, 1992.
Davis, Burke. *Gray Fox.* N.Y.: Holt, Rinehart & Winston, 1956.
Davis, Johnda. *Journal of Jonathan Alder.* Columbus, Ohio: Ohio Historical Society,1988.
Dills, R. S *History of Fayette County.* Evansville, IN: Unigraphic, 1881.
—— *History of Greene County,* 1881.
Downey, Fairfax. *Famous Horses of The Civil War.* N.Y.: Thomas Nelson & Sons, 1960..
Ensminger, M. E. *Horse Husbandry.* Danville IL: Interstate Printers & Pub., 1951.
Evans, Nelson. *History of Adams County.* West Union, Ohio: Emmons Stivers, 1900.
Fansler, Ludene E. "Laurence A. Fansler," *Champaign County, Ohio. 1991.* Dallas, Texas: Taylor Pub. Co.,1991.
Finley, Rev. James B. *Autobiography of Reverend James B. Finley or Pioneer Life in the West.* Cranston & Curts, Cincinnati, 1853.
——. *Life Among The Indians.* Cincinnati, Ohio: Hitchcock & Walden, 1857.
Fourteenth Annual Report of The Ohio State Board of Agriculture 1859. Columbus, Ohio: R. Nevins, State Printers, 1860.
Galloway, William A. *Old Chillicothe.* Xenia, Ohio: Buckeye Press, 1934.
Garrett, Betty. *Columbus, America's Crossroads.* Tulsa, OK: Continental Heritage Press, 1980.
Gibbons, Gail. *Wolves.* N.Y.: Holiday House, 1994.
Gilliland, Thaddeus. *History of Van Wert County.* Chicago: Richland & Arnold, 1906.
Hatcher, Harlan. *Buckeye Country.* N.Y.: G. P. Putman Sons, 1940.
Hatcher, Harlan, Robert Price, Florence Murdoch, John W. Stockwell, Ophia D. Smith. *Johnny Appleseed: A Voice in The Wilderness.* Cincinnati, Ohio: Johnny Appleseed Memorial Library, 1966.
Havighurst, Walter. River to the West. N. Y.: Penquin Group, 1948.
Hildreth, S. P. *Biographical and Historical Memoirs of The Early Pioneer Settlers of Ohio With Narratives of Incidents and Occurrences in 1775.* Cincinnati, Ohio: Derby & Co., 1852.
History of Columbiana County. Philadelphia, PA: D. W. Ensign & Co., 1879.

History of Hardin County. Chicago: Warner & Beer & Co., 1883.
History of Logan County and Ohio. Chicago: O. L. Baskin, 1880.
History of Marion County. Chicago: Leggett, Conaway & Co., 1883.
History of Monroe County. Ohio. Mt. Vernon, Ohio: Atlas Publishing Co., 1898.
History of Preble County. Cleveland, Ohio: H. Z. Williams & Bros., 1881.
History of Richland County, Ohio. Mansfield, Ohio: Graham & Co., 1880.
History of Van Wert & Mercer Counties. Wapakoneta, Ohio: R. Sutton & Co., 1882.
History of Wyandot County. Chicago: Leggett, Conaway & Co., 1883.
Horwitz, Tony. *Confederate in The Attic.* N.Y.: Pantheon Books, 1998.
Howe, Henry. *Historical Collections of Ohio.* Cincinnati, Ohio: Derby, Bradley & Co., 1848.
—— *Historical Collections of Ohio.* Vol. 1. Norwalk, Ohio: Laning Printing Co., 1896.
—— *Historical Collections of Ohio* Vol. 2. Cincinnati, Ohio: Krehbiel & Co., 1904.
Howells, William Dean. *Stories of Ohio.* Cincinnati, Ohio: American Book Co., 1897.
Hurt, R. Douglas. *The Ohio Frontier.* Bloomington & Indianapolis, IN: Indiana University Press, 1996.
Indians of The Plains. American Heritage Editors. Harlan, IA: American Heritage Pub. Co., 1960.
Jenkins, J. Brian. *Citizen Daniel (1775-1835) and The Call of America.* Hartford, CT: Aardvark Editorial Services, 2000.
Kays, Donald J. *The Horse.* N.Y.: Rinehart & Co., 1953.
Kenton, Edna. *Simon Kenton, His Life and Period 1755-1836.* N.Y.: Random House from the Doubleday edition, 1930.
Larkin, Stillman C. *Pioneer History of Meigs County.* Columbus, Ohio: Berlin Printing Co., 1908.
Mack, Horace. *History of Columbiana County.* Philadelphia, PA: D. W. Ensign Co., 1888.
MacLeod, Normand. *Detroit to Fort Sackville, 1778-1779, The Journal of Normand MacLeod.* Detroit: MI: Wayne State University Press, 1978.
Martin, William T. *Martin's History of Franklin County, Ohio.* Columbus, Ohio: Follett, Foster & Co., 1858.
Mathews, Alfred. *Ohio and Her Western Reserve.* N.Y.: D. Appleton & Co., 1902.
McCabe, James. *Planting The Wilderness.* Boston: Lee & Shepard, 1892.
Medert, Patricia. *Raw Recruits and Bullish Prisoners.* Jackson, Ohio: Jackson Pub. Co., 1992.

Meyer, Richard. *Cemeteries and Gravemarkers.* Logan, Utah: Utah State University Press, 1992.
Miami County History. Miami County Ohio Sesquicentennial Committee. Columbus, Ohio: F. J. Heer Printing Co., 1953.
Michaux, F. A. *Travels to the West of the Allegheny Mountains.* London, 1805. Edited by R. G. Thwaites in *Early Western Travels*, Vol. 3. Cleveland, 1904.
Middleton, Judge Evan P. *History of Champaign County, Ohio.* Indianapolis, IN: Bowan & Co., 1917.
Midwest, Collection From Harper's Magazine. Harper's Magazine Editors. N.Y.: Gallery Books, 1991.
Miller, C. G. *Donn Piatt: His Work and His Ways.* Cincinnati, Ohio: Robert Clarke & Co., 1893.
Miller David. *Practical Horse Farrier.* Hamilton, Ohio: E. Shaeffer, 1830.
Morgan, Ted. *Wilderness at Dawn.* N.Y.: Simon & Schuster, 1993.
Nineteenth Annual Report of The Ohio State Board of Agriculture, 1864. Columbus, Ohio: R. Nevins, State Printers.
Norton, A. Banning. *History of Knox County.* Columbus, Ohio: R. Nevins Printer, 1862.
Ohio Fourteenth Federal Census 1820-1900.
Ohio Frontier: An Anthology of Early Writings. Edited by Emily Foster. Lexington, KY: University of Kentucky Press, 1996.
Parkman, Francis. *The Oregon Trail.* Garden City, N.Y.: Doubleday & Co., 1849.
Peaceful, Leonard. *A Geography of Ohio.* Kent, Ohio: Kent State University Press, 1996.
Peck, Robert McCracken. *Land of The Eagle.* N. Y.: Summit Books, 1990.
Pittenger, Rev. William. *Capturing a Locomotive.* Philadelphia, PA: J. B. Lippincott & Co., 1881.
Price, Robert. *Johnny Appleseed: Man and Myth.* Urbana, Ohio: Urbana University, 1954.
Prince, B. F. *Standard History of Springfield and Clark County, Ohio.* Chicago: American Historical Society, 1922.
Raitz, Karl. *Guide to The National Road.* Baltimore & London: John Hopkins University Press, 1996.
—— *The National Road.*
Report of The Commissioner of Agriculture For The Year 1867. Washington D.C.: Government Printing Office, 1868.
Report of The Commissioner of Patents For The Year 1850. Washington D.C.: Office of Printers to House of Representatives.
Revolutionaries, The. by Editors of Time-Life. Alexandria, VA: Time-Life Books, 1996.

Riker, Ben. *Pony Wagon Town*. Indianapolis, IN: Bobbs-Merrill Co., 1948.
Rockel, William. *20th Century History of Springfield and Clark County, Ohio*. Chicago: Biographical Pub. Co., 1908.
Roseboom, Eugene and Francis Weisenburger. *A History of Ohio*. Columbus, Ohio: F. J. Heer Printing Co., 1934.
Rubio, Manny. *Rattlesnake, Portrait of a Predator*. Washington D.C.: Smithsonian Institution, 1998.
Rust, Orton G. *History of West Central Ohio*. Indianapolis, IN: Historical Pub. Co., 1934.
Ryden, Hope. *Bobcat Year*. N. Y.: Viking Press, 1981.
Sears, A. B. *Thomas Worthington*. Columbus, Ohio: The Ohio State University Press, 1958.
Sibley, Warren D. *History of Woodstock, Ohio*. N. P., 1907.
Siedel, Frank. *The Ohio Story*. Dayton, Ohio: Landfill Press, 1950.
Smith, Col. James. *An Account of The Remarkable Occurrences in The Life and Travels of Col. James Smith, During His Captivity With The Indians, in The Years 1755, '56, '57, '58, and '59*. Cincinnati, Ohio: Robert Clarke & Co., 1870.
Smith, Thomas H. *An Ohio Reader, 1750 to The Civil War*. Grand Rapids, MI: Eerdmans, 1975.
Smith, William and Ophia. *Buckeye Titian*. Cincinnati, Ohio: Historical & Philosophical Soc. of Ohio, 1968.
Steele, William. *Old Wilderness Road, An American Journey*. Orlando, FL: Harcourt, 1968.
Steinhart, Peter. *The Company of Wolves*. N.Y.: Alfred Knopf, 1995.
Stille, Samuel H. *Ohio Builds a Nation*. Chicago: Arlendale Book House, 1962.
Swisher, James. *How I Know*. Cincinnati, Ohio: Press of Jones Bros., 1880.
Ware, Joseph. *History of Mechanicsburg, Ohio*. Columbus, Ohio: F. J. Heer Printing Co., 1917.
Williamson, C. W. *History of Western Ohio and Auglaize County*. Columbus, Ohio: W. M. Linn & Sons, 1905.
Wood, Charles S. *Camp-fires on the Scioto*. Boston & Chicago: W. A. Wilde & Co., 1905.
Yearbook of The United States Department of Agriculture, 1895. Washington D.C.: Government Office Printing, 1896.
Yearbook of The United States Department of Agriculture, 1910. Washington D.C.: Government Office Printing, 1911.
Zeisberger, Rev. David. *David Zeisberger's History of The Northern American Indians*. 1910. Reprint. Edited by Archer Hubert & William Schwarze. Lewisburg, PA: Wennawoods Pub., 1999.

MAGAZINES:
Barsness, Larry. "Piskiou, Vaches, Savages, Buffer, Prairie Beeves — Buffalo." *American Heritage,* Oct.-Nov., 1979, p 25-34.
Country Kids. (N. D.).
Echoes. Vol. 39, No. 5, Oct./Nov., 2000, p 3.
Gilbert, Bil. "Pioneers Made a Lasting Impression on Their Way West." *Smithsonian,* May, 1994,p 44.
Habverson, Deborah. "Journey of Jonathan Hale." *Early American Life,* June, 1980, p 20-76.
"How Leather is Tanned." *American Agriculturist,* Apr., 1874.
Hurt, R. Douglas. "Bettering the Beef." *Timeline,* Mar/Apr, 1993, p 25.
McMillan, Jean. "Family Papers Reveal History of The Hunts." *Echoes,* Oct./Nov., 2000.
Rupp, Rebecca. "Animal Friends." *Early American Homes,* Feb., 1997.
Terry, Karen. "Passion For Beaver Pelts. *Cobblestone,* Dec., 1991.
Young, James. "For Man And Beast. *Timeline,* Ohio Historical Society publication, Sept/Oct., 2001.

MISCELLANEOUS:
Banes, Margaret Ward. obituary, 1890.
Bradley, Cyrus P. "Journal of Cyrus P. Bradley, A Trip Through Ohio and Michigan in 1835." Ohio State Archaeological and Historical Society Quarterly, 1906.
Carlo, Edwin. letter. St. Paris, Ohio, circa 1870.
Estes, Ralph S. ed. *Ohio Historical Review* 3, 9.
Everhart, Warren. "Jezebel." circa 1956. mimeographed.
Glessner, Anna. letter to Champaign County Historical Society re Rev. Merrill's "Ox Sermon," Jan. 31, 1939.
James, John H. "Reminiscences of James Taylor." (N. D.) mimeographed.
Knight, William J. "Some of The Early Campmeetings in Champaign County," circa 1906. mimeographed.
Ordinances of The Town of Urbana, Ohio, March, 1826-July 31, 1867.
Ross, R. H. Champaign County Historical Society Museum, circa 1943. mimeographed.
Ward, Elizabeth Hughes. 1814-1838 diaries. Champaign County Historical Museum.

ABOUT THE AUTHOR

Barbara Stickley Sour is a member of the Ohio Historical Society and served several terms as president of the Champaign County Historical Society. She currently is on the Champaign County Historical Society board of trustees and is editor of the monthly newsletter. She also sits on the board of directors for the Champaign County Preservation Alliance. Under the auspices of the Champaign County Historical Society she wrote *A Tour Through Champaign County, Ohio* which includes illustrations, photographs, and descriptive information about pre-1900 buildings.

www.ingramcontent.com/pod-product-compliance
Lightning Source LLC
Chambersburg PA
CBHW051627230426
43669CB00013B/2214